MODELS OF DECISION-MAKING

Classical decision theory evaluates entire worlds, specified so as to include everything a decision-maker cares about. Thus, applying decision theory requires performing computations far beyond an ordinary decision-maker's ability. In this book Paul Weirich explains how individuals can simplify and streamline their choices. He shows how different "parts" of options' outcomes (intrinsic, temporal, spatiotemporal, causal) are separable, so that we can know what difference one part makes to the value of an option, regardless of what happens in the other parts. He suggests that the primary value of options is found in basic intrinsic attitudes toward outcomes: desires, aversions, or indifferences. And using these two facts he argues that we need only compare small parts of the outcomes of options we face to make a rational decision. This important book will interest readers in decision theory, economics, and the behavioral sciences.

PAUL WEIRICH is a Curators' Professor in the Philosophy Department at the University of Missouri. His previous books include *Collective Rationality* (2010), *Realistic Decision Theory* (2004), *Decision Space* (Cambridge, 2001), and *Equilibrium and Rationality* (Cambridge, 1998).

MODELS OF DECISION-MAKING

Simplifying Choices

PAUL WEIRICH
University of Missouri

CAMBRIDGE
UNIVERSITY PRESS

CAMBRIDGE
UNIVERSITY PRESS

University Printing House, Cambridge CB2 8BS, United Kingdom

One Liberty Plaza, 20th Floor, New York, NY 10006, USA

477 Williamstown Road, Port Melbourne, VIC 3207, Australia

4843/24, 2nd Floor, Ansari Road, Daryaganj, Delhi - 110002, India

79 Anson Road, #06-04/06, Singapore 079906

Cambridge University Press is part of the University of Cambridge.

It furthers the University's mission by disseminating knowledge in the pursuit of education, learning and research at the highest international levels of excellence.

www.cambridge.org
Information on this title: www.cambridge.org/9781107434783

First published 2015
First paperback edition 2017

A catalogue record for this publication is available from the British Library

Library of Congress Cataloging in Publication data
Weirich, Paul, 1946–
Models of decision-making : simplifying choices / Paul Weirich.
pages cm
ISBN 978-1-107-07779-9 (hardback)
1. Decision making. I. Title.
QA279.4.W453 2015
519.5′42–dc23
2014027896

ISBN 978-1-107-07779-9 Hardback
ISBN 978-1-107-43478-3 Paperback

For my teachers

Contents

Figures

Tables

Preface

Rational decisions rest on evaluations of options, and an option's evaluation in ideal cases surveys all considerations that count for or against the option. In ideal cases, with unlimited computational capacity, an option's evaluation may, without harm, process irrelevant considerations along with relevant ones. However, in real cases, where efficiency matters, an optimal evaluation surveys only relevant considerations. This book identifies the relevant considerations to which an optimal evaluation attends.

The book's approach to choice is philosophical. Its principles evaluate decisions for rationality. Because efficient methods of evaluation improve deliberations about options, they promote practicality. Besides students and specialists in philosophical decision theory, scholars in all fields that draw on accounts of rational decisions will be interested in efficient evaluation of options.

For financial support, I thank the University of Missouri Research Council, Research Board, Faculty Development Program, and Center for Arts and Humanities. For congenial study quarters, I thank the University of Pittsburgh Center for Philosophy of Science, the University of Sydney Centre for the Foundations of Science, and the Australian National University School of Philosophy. For expert advice, I thank the manuscript's anonymous readers. Notes give credit to many people who helped with various sections. Finally, many thanks to my excellent editorial team.

Introduction

A typical exercise in a textbook on rational decisions imagines a diner choosing a meal from the menu at a restaurant. For simplicity, it supposes that the diner ranks items on the menu using only her tastes at the moment. However, a reasonable diner considers her decision problem's spatiotemporal context. She may consider the provenance of the restaurant's food and so look toward the past. She may also consider the consequences of her choice for causes she supports, such as local farming, and so look toward the future.

To make sure that a decision's evaluation considers all relevant factors, not just events here and now, some principles of evaluation target the *possible world* that results given the decision's realization.[1] The possible world covers the decision's past, present, and future. It has spatiotemporal structure and includes all the events that occur given the decision's realization. Evaluating it includes the consideration of every relevant pro or con no matter how distant in space or time.

Comprehensively evaluating a decision by evaluating the decision's world is safe but has an obvious drawback – complexity. It makes appraisal of ordering a dish from a menu an enormously complex appraisal of the possible world issuing from the order. The possible world has so much detail that no description, no matter how elaborate, comes close to being thorough. Practicality demands simplification.

This book shows how to narrow the scope of a decision's evaluation. An evaluation of a choice from a menu need not survey all events in the decision's world, including events in the remote past, such as the burning of the library of Alexandria in 48 B.C. Although the diner cares about that calamity, evaluation of ordering a ragout tonight may ignore it.

---

[1] I italicize the first occurrence of a technical term in a passage that introduces it. Usage of technical terms in decision theory is not uniform, and I introduce technical terms to suit the principles I formulate with them. Each principle's formulation reviews the points motivating its terminology.

An efficient evaluation trims three types of irrelevant consideration. It puts aside, first, matters of indifference; second, nonbasic sources of utility; and, third, unavoidable events, that is, events that occur given every option available. The first and second steps narrow considerations enough to make them describable. The third trims the past, for example. The evaluation's elaboration occupies most of the book because decision theory needs new resources to justify principles that ignore sources of utility.

People manage to focus on information relevant to tasks at hand, although their methods are sometimes opaque. For example, people recognize faces more easily than do computers because people pick out relevant facial features in ways not yet formalized. Although people generally focus on relevant considerations when evaluating options, their methods lack precise description and justification. The following chapters explicate and justify some common methods of identifying relevant features of options in a decision problem; in particular, methods that focus on an option's future, its spatiotemporal region of influence, or its consequences. Rather than offer new practical advice, they make precise familiar methods of simplifying choices and give them firm philosophical foundations.

Principles

Normative decision theory formulates principles for evaluating the options in a decision problem and for identifying rational options. Taking the theory's perspective, a feature of an option is relevant to its evaluation if the feature has a role in identifying rational options. Rationality as it applies to options is a source of standards for a decision among options. Some texts adopt technical definitions of rationality according to which, for example, an option is rational if and only if it is at the top of a preference ranking of options. Such technical definitions suit certain projects, but this book treats rationality in its ordinary evaluative sense and so grounds principles of rationality in normative arguments rather than technical definitions. Its arguments draw on judgments about rationality in particular cases and judgments about rationality's general features.

An adequate representation of an agent's decision problem specifies an exhaustive set of options available at the time the decision problem arises. Comparisons of options in a decision problem establish which options are better than others, and which are best, according to the agent's goals. These comparisons ground rational decisions. For simplicity, assume that an assignment of utilities to options settles the preference ranking of the options in a decision problem. Then, the key to a rational decision is the assignment of utilities to options.[2]

The principle of utility maximization states that a rational decision maximizes utility. This means that the decision's utility is at least as great as any alternative decision's utility. The principle's restriction to decision problems in which some decision maximizes utility ensures the possibility of compliance. Its restriction to ideal agents in ideal circumstances makes compliance feasible. The principle is too demanding for real agents with cognitive limits and time pressure.

The principle of utility maximization compares options by assigning a *utility* to each option. I take an option's utility to indicate the agent's strength of desire to realize the option. Strength of desire depends on the agent's information. If she believes that an option will have only good consequences, she desires to realize it to some extent; but if she learns that it will have only bad consequences, she then becomes averse to realizing it. Because the principle that a rational choice maximizes utility takes rationality in its ordinary evaluative sense and defines utility in terms of desires rather than choices, it is a normative principle rather than a definitional truth.

By taking utility as strength of desire, so that an option's utility is not relative to a set of preferences, I adopt a *realist* view of utility that contrasts with a *constructivist* or *interpretativist* view that extracts utility from a set of preferences. For example, the theory of *revealed preference*, common in economics, takes a constructivist view of utility. It uses utility just to represent preferences. The book's methods of simplifying choices work given a constructivist conception of utility; however, adopting a realist conception of utility strengthens normative principles of utility. Chapter 2 reviews the reasons for the realist view.

The goal of blending probability and utility in principles of choice suggests taking both probability and utility, and not just probability, to attach to a

---

[2] The assumption of quantitative evaluations of options as a basis for a decision is not significantly stronger than the assumption of comparisons of options. Although identifying a top-ranked option relies directly on a ranking of options and only indirectly on quantitative evaluations of options, ranking probability mixtures of options requires quantitative evaluations of the options mixed. Granting that probability mixtures of options are options, ranking options requires quantitative evaluations of options. In any case, assuming quantitative evaluations of options does not significantly affect conclusions about efficiency in deliberations. Because deliberations use options' evaluations to compare options, reasons to ignore considerations when evaluating an option (such as considerations that apply to all options) resemble reasons to ignore considerations when comparing two options (such as considerations that apply to both options). Identification of relevant considerations is similar for evaluation of options and for comparison of options.

proposition, the content of a declarative sentence. Desire may be taken as an attitude toward a proposition. A sentence stating a proposition that an agent wants to be true expresses a target of the agent's desires. For example, an agent may desire that peace prevail. Then, an object of the agent's desires is the proposition that peace prevail. Desiring that an event occur is equivalent to wanting that the event occur. This sense of desire, which Schroeder (2004) expounds, is common in philosophy, and Chapter 2 introduces it more thoroughly. Taking desire as a propositional attitude, an option's utility attaches to a propositional representation of the option.

Besides depending on information, an option's utility depends on the scope of the option's evaluation. It may consider, for example, just the logical consequences of the option's realization, the total consequences of its realization, or the possible world the option realizes. To be comprehensive, ordinary utility evaluates the world that would be realized if the option were realized. When that world is uncertain, the option's utility equals a probability-weighted average of the worlds that might be realized if the option were realized.

This book adopts *causal decision theory's* analysis of an option's utility without reviewing the arguments for it.[3] According to one version of this analysis, namely, *world-Bayesianism* as Sobel (1994: chap. 1) presents it, a decision's utility is its world's utility, or if that is unknown, a probability-weighted average of the worlds that, epistemically speaking, might be its world. A decision's world is the possible world that *would* be realized if the decision *were* made. According to a standard account of conditionals, this possible world, among all worlds containing the decision, is nearest to the decision problem's world, assuming, for simplicity, that a unique world is nearest. Introducing the decision's world using a conditional in the subjunctive mood signals that causal relations settle the nearest world with the decision's realization, as Chapter 2 explains.

An initial simplification of a decision's world puts aside matters of indifference. It takes a decision's world as a proposition specifying for each desire and aversion whether it is realized. A further simplification excises other irrelevant parts of the decision's world, for example, the decision's past. These simplifications reduce an evaluation's information-processing demands.

World-Bayesianism advances a standard of evaluation for options, not a procedure for deliberation. An outsider whose cognitive powers differ from the agent's cognitive powers may apply the standard of evaluation. Efficient

---

[3] Weirich (2012) reviews the arguments.

principles for evaluation of options, even taken as standards of evaluation rather than as procedures for deliberation, ask too much of cognitively limited agents, especially in adverse circumstances, but make appropriate demands of ideal agents in ideal circumstances. Ideally situated ideal agents may convert principles of evaluation into procedures of deliberation and therefore lack excuses for failing to satisfy the principles of evaluation. Although I advance principles of utility as standards of evaluation because I restrict the principles to ideally situated ideal agents, the principles also yield procedures of deliberation.

Utility analysis, because it investigates rationality for an ideally situated ideal agent, treats a normative *model*. The model is a possible world like the actual world except for including an ideal agent in an ideal decision problem. Utility analysis fills out this sketch of the model by specifying respects in which an agent and a decision problem are ideal. The model is not mathematical, although some principles of utility are mathematical. The principles, which evaluate options to identify rational options, are not part of the model because, being necessary truths, they govern all possible worlds without distinguishing the possible world the model presents. Consequently, the model is normative because of its function rather than because it has normative components. The model does not trivialize the arguments for the principles of utility, although it removes objections to the principles. Ideal agents in ideal decision problems can violate the principles, although the arguments maintain that these agents, if rational, observe the principles.

Although realism prompts an interest in efficiency, methods of achieving efficiency may incorporate some idealizations. A method may reduce information-processing demands while idealizations accompanying it ensure enough cognitive power to meet its remaining demands.

The book's efficient decision principles target ideal agents who calculate effortlessly and know all logical truths. After evaluating options, such agents can compare options and decide without any cognitive costs. However, they can only up to a limit costlessly process considerations to evaluate an option. Efficient evaluation reduces the number of considerations to process so that evaluating an option does not exceed their limit. Because such agents, although cognitively powerful, beyond their limit pay evaluation costs, an efficient method of evaluating options benefits them. Its availability removes an excuse for failing to maximize utility and justifies holding them accountable for utility maximization.

Selective adoption of idealizations assists progress toward a theory of rationality for real agents. Idealizations simplify principles of rational choice, and future research may remove the idealizations to generalize the principles. Weirich (2004) shows how to remove, from a normative model, idealizations such as the assumption that agents have precise probability and utility assignments.[4] Until theorists generalize idealized principles so that they also cover nonideal cases, the principles are just helpful guides in real-life decisions. The idealized principles express a goal that rational agents in real cases aspire to attain, they direct consultants and computers offering aid with decisions, and they apply approximately to some real cases. Also, some implications of the principles hold robustly in the absence of idealizations. For example, in a typical decision problem, a person evaluating an option does well to put aside past events and examine only the option's consequences, even if the shortcut's precise justification depends on idealizations. In particular, the advice is apt for a diner ordering from a menu, even if its incorporation by the directive to maximize utility rests on idealizations.

Efficiency also has theoretical interest apart from its practical interest. Identifying the considerations relevant to an option's evaluation advances a theory of rationality by refining explanations of a decision's rationality. An explanation improves by using only considerations that bear on a decision's rationality. A completely satisfactory explanation, for economy, ignores irrelevant considerations. A good explanation of the rationality of a diner's decision from a menu skips fires in ancient Alexandria, for instance. Efficient decision principles ground good explanations of a decision's rationality.

Efficiency

Although explaining efficiency in evaluation of options does not require a thorough account of efficiency, a brief account clarifies the project. This section defines the type of efficiency the principles of utility analysis achieve. It paints the background for the claim that evaluating options according to their futures is more efficient than evaluating them according to their worlds.

Many types of efficiency are valuable. I examine *efficiency in doing a task*. Specifying this type of efficiency involves specifying the task – including the

---

[4] A model with idealizations may offer a partial explanation of a target phenomenon. Consider the explanation of a person's checkbook balance. Suppose that a model contains checks and deposits but omits fees the bank levies. The model offers a partial explanation of the balance by displaying the operation of some factors affecting the target phenomenon. A full explanation adds the fees that the model omits. A model of a decision's rationality may similarly offer a partial explanation of its rationality. A full explanation treats factors that idealizations put aside.

resources available, the working conditions, and the final product. It also involves specifying the measure of efficiency, which may target just a part of overall efficiency.

A method of doing a task may be efficient without being maximally efficient; an efficient process need not be the most efficient process. In a classificatory sense, an efficient process completes the task without wasting resources; it uses close to the minimum required. Comparative efficiency ranks methods; some are more efficient than are others. Comparisons are transitive and in finite cases identify the most efficient processes. A method of performing a task achieves maximum efficiency if and only if no alternative method is more efficient.

Performing a task efficiently may differ from performing it in a good way because efficiency, although good, differs from overall goodness. A method of production may be efficient but not good because its product is worthless. Also, a process may be efficient in some respect even if it is not efficient overall; it may achieve one type of efficiency at the expense of another more important type of efficiency. Overall efficiency weighs and combines types of efficiency to find a suitable balance among them. To simplify, a study of efficiency may treat just one component of overall efficiency.

Efficiency in a task obtains relative to a measure of performance. For travel from Chicago to New York, a driver may evaluate routes according to the number of miles or the number of hours behind the wheel. A factory may measure efficiency in an automobile's production using a combination of raw materials, time, and costs (perhaps including environmental impact), or it may use profit. In computing, performing the task with the fewest steps or in the shortest time is a convenient measure of efficiency. Another measure may use the length of the program for the task. In modal logical, a measure of efficiency for an axiomatization may use the number of symbols in the axiomatization. An alternative measure may use the lengths of proofs of a representative set of theorems using the axiomatization and a set of rules of proof.

Efficiency in a task depends on a precise specification of the task. The task may be general, such as building automobiles, or it may be a specific instance of a general task, such as building a particular automobile. Specifying the resources, conditions, and product is a way of specifying the task. Resources for building automobiles may include not only metal used to make a chassis but also workers and tools. Specifying the task also includes specifying the product. It may be an automobile that meets high standards of quality. Efficiency evaluates a process going from a specific starting point to a specific end point in a specific environment.

These points about efficiency in a task carry over to efficiency in evaluation of an option. The task is evaluation of the option for the purpose of
choosing an option. The resources include an ideal agent's cognitive abilities; an identification of basic intrinsic desires and aversions, or *basic
intrinsic attitudes*, that the option's world realizes; and their realizations'
temporal, spatiotemporal, and dependency relations to the option. The
conditions include the decision problem's being ideal so that probabilities
and utilities of the option's possible outcomes are available. The product is
an accurate evaluation of the option, namely, the option's utility. The
measure of efficiency is the number of basic intrinsic attitudes reviewed.

Identifying the general method of evaluation that processes the fewest
considerations is just a step toward overall efficiency. Simplifying choices
requires many types of efficiency, and calculations of overall efficiency
weigh processing just relevant considerations along with other types of
efficiency. Additional types of efficiency may, for example, count the
calculations an evaluation performs.

Because deliberation in a decision problem includes, besides evaluation
of options, identification and selection of options, efficient evaluation does
not entail efficient deliberation. Efficient deliberation may not use efficient
evaluation if identifying relevant considerations is costly. Deliberation that
sorts considerations by relevance may be inefficient. However, given the
sorting as a resource, an evaluation that processes only the relevant considerations is efficient.

Given the uncertainty of an option's world, an option's evaluation uses
probabilities of worlds that might be the option's world. In a broad sense,
relevant considerations include features of evidence that influence these
probabilities. However, given the probabilities as a resource, an efficient
evaluation does not process the evidential considerations. An efficient
evaluation, granting my specification of the task, processes only, and not
necessarily all, considerations relevant to evaluating the worlds that might
be the option's world.

An efficient principle of utility maximization minimizes a suitable utility
assignment's demands on an agent's information about options; it efficiently obtains options' utilities. Its utility assignment uses only information necessary for generally applying the principle of utility maximization.
Narrowing the scope of an option's evaluation achieves this efficiency. The
narrower the scope, the fewer considerations the evaluation reviews and
hence the lower its cognitive cost, other things being equal.

Efficient general principles for decisions use a minimum of the agent's
information about options sufficient in every decision problem for resolving

the problem. Specialized principles may further minimize informational demands because the demands vary from problem to problem. Suppose that in a decision problem, an option realizes some aversions and no desires, and a second option realizes some desires and no aversions. The second option is superior to the first option no matter which aversions the first option realizes and which desires the second option realizes. In this case, comparison of the options need not review the two options' consequences. Assuming that the rational options a decision problem's resolution identifies are at the top of a preference ranking of options, an efficient principle for the decision problem may compare the two options without placing each option on a utility scale according to its consequences. Evaluation by consequences, although generally efficient, is not in every decision problem a maximally efficient way of ranking options.

For perspective, consider an efficient way to select three green marbles from a bag. A straightforward method draws marbles one by one, inspects each, and keeps it if it is green, until collecting three green marbles. Another method takes three marbles at once and, if necessary, replaces marbles not green to obtain three green marbles. The second method is more efficient if all the marbles in the bag are green. It takes fewer steps handling individual marbles in that case. Efficiency depends on circumstances.

Efficient general methods of evaluation make informational demands no greater than other general methods of evaluation. In special cases, some decision principles may be more efficient with respect to information processing than an efficient general principle. In case-by-case comparisons, one decision principle may not be uniformly more efficient than another but may as a general principle nonetheless be more efficient than the other principle because its general application requires less information than does the other principle's general application. This book refines for efficiency the principle of utility maximization. Although the refined principle is not efficient in every decision problem, it achieves efficiency with respect to other general decision principles. It uses an efficient general method of assigning utilities to the options in a decision problem to rank options and thereby resolve the decision problem.

Characterizing the relevant considerations for comparing options is a theoretical project. The characterization also has practical value because it licenses ignoring parts of an option's world to make choice efficient. However, theoretically efficient and practically efficient evaluations may differ. In some cases, theoretically efficient evaluations do not achieve practical efficiency because identifying relevant considerations requires reviewing all considerations anyway. Applying a method of evaluation

that is optimal in theory may not be optimal in practice because the method
processes hard-to-identify considerations. A theoretically efficient evaluation procedure may not be practically efficient because of its application's
cognitive demands.[5]

To illustrate this point about efficiency, consider an example from
computer science. Sometimes efficient methods of computation are hard
to implement. To sort n items, an efficient method uses $n \times \log(n)$ steps. To
sort 1000 items, it takes $1000 \times \log(1000)$ steps, or 3000 steps. The efficient
method is hard to implement. So, one might use an easy implementation of
an inefficient process that uses n-squared steps. Although it takes 100,000
steps to sort 1000 items, it may be more practically efficient than its
theoretically efficient rival. Similarly, an efficient decision method may
consider only options' consequences, although an easy-to-implement, inefficient method may consider options' worlds. Theoretical efficiency yields
practical efficiency only given idealizations that facilitate implementation.

In some cases, a theoretically simple evaluation of an option is impractical
because the evaluation must review every consideration to ascertain whether
it is relevant. In other cases, however, the evaluation may dismiss irrelevant
considerations in batches so that it is practical. For example, it may dismiss
the past without reviewing every past event. Although theoretically efficient
evaluations are sometimes cognitively demanding, they have practical value
in cases where people can perform them.

Some reasonable simplifications of evaluations risk failure to identify a
utility-maximizing choice. Deliberation costs may justify evaluating options
with respect to a single objective, such as money, although the simplification neglects relevant considerations. Reasonable simplification continues
beyond the least taxing evaluations that identify a utility-maximizing
choice. It trades accuracy in resolving decision problems for reduced cognitive costs. This book's assessment of decision methods for efficiency
ignores such trade-offs and compares only accurate methods of resolving
decision problems.[6]

---

[5] Shanahan (2009) states artificial intelligence's frame problem in its general form: "How do we account
for our apparent ability to make decisions on the basis only of what is relevant to an ongoing situation
without having explicitly to consider all that is not relevant?" Overall efficiency for deliberations has
the goal of identifying relevant considerations without reviewing all considerations and so the goal of
solving a philosophical version of the frame problem.

[6] Gigerenzer (2000: chap. 8) describes decision heuristics that promote efficiency in realistic environ-
ments, but at the cost of accuracy in general applications. For example, the heuristic to "Take the Best
and Forget the Rest" makes a decision using one type of reason and ignoring other types of reason.
Although it efficiently yields good decisions in a congenial environment, it fails when the reasons
ignored contravene the reasons considered. Utility analysis advances precise general methods of

A theoretically efficient general method of accurately evaluating options therefore makes three concessions. First, a method designed for a special case may be more efficient than the general method in the special case. Second, practical problems may hinder the method's implementation; its efficiency supposes idealizations that facilitate evaluations achievable in theory. Third, sacrificing accuracy may reasonably achieve greater overall deliberational efficiency than the method achieves.

Efficiency compares methods of performing a task with specified resources in specified circumstances. This book treats efficiency for general methods of evaluating options in a decision problem using probability and utility assignments for the options' realizations of basic intrinsic desires and aversions, given that conditions are ideal for applying the methods. It puts aside other types of efficiency, such as conceptual economy in the definition of an option's utility. The number of basic intrinsic desires and aversions an option's utility assignment considers supplies the measure of efficiency; the fewer considerations an option's utility assignment processes, the more efficient it is. This operational efficiency applies to an option's evaluation given sorted considerations to process. Identifying evaluations with this circumscribed type of efficiency assists more general treatments of efficiency that consider overall cognitive costs and trade-offs with accuracy, given an agent's limited resources for evaluating options. Establishing that an accurate evaluation of options need not process, for example, past realizations of desires points the way toward overall efficiency in deliberations.

Theoretical issues

Twentieth-century decision theory has *operationalist* foundations, and some recent works, such as Gilboa (2009), retain these foundations. Operationalism imposes standards for the introduction of theoretical terms. In the empirical sciences, confirmation of hypotheses containing theoretical terms generally requires testing that demands operational methods of applying the terms. Support for normative principles, however, does not demand operational application of their theoretical terms. The case for operationalism thus is stronger in the decision sciences than in normative decision theory. Operationalism has various formulations. Some formulations merely advocate methods of measuring theoretical entities, properties,

evaluating options that achieve efficiency without sacrificing accuracy. A justification of heuristics that suit a person's ordinary decision problems explains why circumstances warrant departures from these precise general methods.

and relations. These formulations are reasonable. Other formulations prohibit concepts not defined in terms of observable operations. These formulations are unjustifiably restrictive if the standard of definition is philosophically strict. Their application fragments a theory by dividing each theoretical entity into many, one for each method of detection, and does not acknowledge the possibility of measurement errors.[7] Chapter 2's introduction of probability and utility describes methods of measuring probability and utility, without using the methods to define probability and utility, and elaborates reasons against applying semantic operationalism to probability and utility. To gain explanatory power, the normative decision theory it presents puts aside operationalist criteria of meaning and so does not require operational definitions of probability and utility. This book uses nonoperationalist conceptual resources, such as utilities defined using desire rather than choice, to enrich its principles and to make them more versatile. The types of utility it introduces meet reasonable standards of meaningfulness and measurability.

For efficiency, an option's evaluation may divide an option's world into two parts so that a single part's utility suffices for the option's evaluation. For example, a division of an option's world into parts using time as a guide may divide the world into the part that comes before the option's realization and the other part, which for brevity I call the option's future. For the division to work, the parts must be *separable* with respect to utility. That is, the utility of the world must be a function, such as a sum, of the utilities of its parts, so that the order of options with respect to a part, when the other part has constant utility, agrees with the order of worlds. Then, granting that the utility of the past is the same for all options, a comparison of options according to their futures is equivalent to a comparison of options according to their worlds. An option's evaluation may omit an evaluation of the option's past and focus on the option's future to reduce information processing. An evaluator's cognitive limits may make it impractical to evaluate an option's world but practical to evaluate its future.

In general, efficiency suggests dividing an option's world into parts that are separable with respect to utility and such that some parts are common to the worlds of all the options available in a decision problem. An evaluation, to reduce information processing, then omits the worlds' common parts. The following chapters explicate and justify common methods of evaluating

---

[7] Chang (2004: 197–219; 2009) rejects operationalism as a theory of meaning but recommends using methods of measurement to ground theoretical concepts, such as temperature. Weirich (2001: sec. 1.4) advances a similar view in decision theory.

wholes by evaluating parts. They supplement standard principles of utility with new principles concerning addition of utilities. The new principles justify an evaluation's trimming considerations.

It may seem that the arrangement of a world's parts, not just their utilities, influences the world's utility. However, a world's parts come with the arrangement they have in the world. The danger facing a division of a world's utility into parts is not that it neglects the parts' arrangement but rather that it omits or double counts relevant considerations. An evaluation that targets an option's future and omits its past must neither omit nor double count a consideration concerning the past that is relevant to the evaluation. It must register just once each relevant influence of the past on the future. For example, wearing caps and gowns to graduation continues a tradition. The tradition's existence influences the consequences of current acts. The past makes some current acts have the consequence of continuing a tradition. An evaluation of wearing caps and gowns omits that consequence if it considers only present convenience. It double counts the consideration if it considers both graduation tradition and commencement tradition, given that these are the same tradition.

Decision theory has not thoroughly argued for efficiency in the evaluation of options. One explanation of the argumentation's absence is the view that common efficiencies, such as ignoring the past, do not need justification. However, philosophical positions such as value holism challenge these common efficiencies. So a decision theory that incorporates them must defend them. Another explanation of the argumentation's absence is the view that holism about the utility of an option's world thwarts evaluations of options that use fragments of an option's world, such as the option's future. According to this view, decision theory correctly forgoes shortcut evaluations. However, this dismissal of the common efficiencies is too quick.

Value holism states that the value of a whole does not divide into the value of its parts. Its slogan is that the whole does not equal the sum of its parts.[8] That the value of a composite may depend on how its parts fit together suggests holism. However, value holism applies relative to a specification of

---

[8] Brown (2007) distinguishes two kinds of value holism. One denies value's additivity, and the other denies value's context-independence, that is, denies its invariability or unconditionality. Value holism in the second sense claims that value is variable. Letting the values of parts change with the wholes to which they belong may support value's additivity. So holism that asserts variability may undermine holism that denies additivity. As Hurka (1998) explains, G. E. Moore, for intrinsic value, denies additivity but affirms invariability. Brown (2007) calls atomism the combination of additivism and invariabilism. Others call invariabilism alone atomism.

a whole, its parts, and a type of value. For a type of utility and a division of a whole into parts, holism errs if the whole's utility equals the sum of the parts' utilities. Although holism is right about some types of utility and some methods of splitting some wholes into parts, a world's utility divides into its parts' utilities, given suitable parts and a suitable type of utility for the parts. This book's methods of analysis, which run contrary to holism, simplify utility calculations and hence identification of rational decisions. Strong justifications support the methods.

How does utility analysis fit with decision theory as Savage ([1954] 1972) formulates it? He acknowledges the possibility of constraints on preferences in addition to the constraints he formulates (59–60). Utility analysis strengthens his theory by advancing additional constraints on preferences. Although he adopts a constructivist view of probability and utility, whereas utility analysis adopts a realist view, moving to the realist view just strengthens his decision principles. In general, utility analysis makes his theory more powerful.

Furthermore, Savage (82–91) notes that his theory's precise application requires taking an option's outcome as a grand world that specifies everything that matters to the agent. Practicality counts against considering grand worlds when resolving a decision problem, and so he suggests considering instead simplified, small worlds. For efficiency, deliberations may replace grand worlds with small worlds that omit some relevant considerations, although this jeopardizes accuracy. For example, buying a lottery ticket with a car as a prize has possession of the car as the outcome given the ticket's selection. This small world outcome does not include the price of gasoline, although the price affects the car's utility. Savage does not present a justification, besides practicality, for relying on small worlds that ignore considerations an agent cares about when they affect a decision problem's resolution. Utility analysis shows that using certain types of small worlds to simplify choices does not compromise accuracy. An option's evaluation may replace an option's world with the set of basic intrinsic attitudes that the option realizes. This reduction, following Savage, attends to an option's consequences, but only selected consequences, namely, realizations of basic intrinsic attitudes.

Preview

Separation of considerations is utility theory's tool for analyzing an option's utility. Evaluations of options that examine multiple attributes of the option's realization consider whether the attributes are separable. For example, evaluation of a proposed storage facility for nuclear waste considers the facility's cost, longevity, and safety. The facility's evaluation depends on whether the considerations are separable or instead relevantly interact, as they do if at low cost longevity matters less than it does at high cost. Whether, for an agent, considerations are separable is an empirical question. For an agent of suitable abilities, some considerations should be separable; norms of separability apply. Principles of rationality may also permit but not require separation of considerations.

The separability of considerations grounds an efficient evaluation of options that simplifies choices. The expected-utility principle handles uncertainty by separating chances for outcomes and as a by-product justifies putting aside chances that all options produce. Basic intrinsic attitudes ground a way of separating the considerations that settle the utility of an option's world. The separation justifies putting aside realizations of basic intrinsic attitudes that occur in every option's world. Chapters 1–3 present separability and establish separability of chances and separability of realizations of basic intrinsic attitudes. Chapters 4–6 use these types of separability to simplify choices in progressively more efficient ways. Evaluation of options may put aside the past, events that no option can influence, and in fact all events except the consequences of options. To evaluate an option, an ideal agent in an ideal decision problem needs only a probability-weighted sum of the intrinsic utilities of the basic intrinsic attitudes that an option's world realizes; moreover, this sum can be cut down to just basic intrinsic attitudes realized in the future, or in a certain spatiotemporal region of the future, or whose realizations are the option's consequences.

Chapters 1–3, as a necessary preliminary, introduce technical ideas that the later chapters use. Chapter 1 provides a general account of separability and a general method of establishing principles governing separation of considerations. The following two chapters justify methods of separating an option's utility into parts.

Chapter 2 treats separation of considerations according to possibilities. A common assessment of a bet separates considerations into the possibility of winning and the possibility of losing. It calculates the bet's *expected utility* to handle uncertainty about the bet's outcome. An option's expected utility divides the option's utility into utilities deriving from the option's possible outcomes. World-Bayesianism, for rational ideal agents, takes an option's utility to equal the option's expected utility, a probability-weighted average of the utilities of the worlds that might be the option's world. Expected-utility analysis separates an option's utility into utilities of chances for its possible outcomes. The chapter precisely formulates and then justifies this type of separation of considerations. The expected-utility principle it formulates applies to all the types of utility that later chapters introduce.

People commonly evaluate an option by considering the goals the option will promote or frustrate. A dean may evaluate a policy for appointing postdoctoral fellows by considering how it promotes the goal of encouraging research and also how it frustrates the goal of trimming the college's budget. Chapter 3 explicates evaluation of options according to goals and aversions. It presents, as a foundation for efficient evaluation of options, intrinsic-utility analysis, which divides a world's utility according to intrinsic desires and intrinsic aversions that the world realizes. It treats, in particular, the realization of *basic intrinsic attitudes*, that is, intrinsic attitudes for which no other intrinsic attitude furnishes a reason, such as desires for pleasure and aversions to pain. They are ultimate desires and aversions rather than instrumental desires and aversions. According to hedonism, a world's utility for an agent is a sum of the utilities of the pleasures and pains the agent experiences in the world. Intrinsic-utility analysis recognizes basic intrinsic desires and aversions that extend beyond pleasures and pains, and it makes precise methods of using them to separate a world's utility into parts. In a decision problem for an agent, intrinsic-utility analysis divides the utility of an option's outcome, a possible world, into the intrinsic utilities of realizations of basic intrinsic attitudes. Chapter 3 uses basic intrinsic attitudes to analyze a world's utility and an option's utility.

Intrinsic-utility analysis is a descendant of *multi-attribute–utility theory*, which divides an outcome's utility into the utilities of its attributes. A variant of multi-attribute–utility theory, *hedonic pricing*, uses features of an object, such as a house, to explain its price, a rough measure of its utility. Multi-attribute–utility theory and hedonic pricing divide considerations but do not advance norms of separability to ground the division. Chapter 3 argues for norms of separability involving basic intrinsic attitudes to ground a division of a world's utility into parts.

Many methods of putting aside considerations fail. For instance, the method of putting aside all past events fails for two reasons. First, the combination of a past event and a future event is not a past event, and so the method does not put aside the combination. By not putting it aside, it fails to ignore the past event that the combination includes. Although the method aims to put aside past events, it entertains them as components of combinations. It is ineffective. Second, being the first to climb a mountain depends on the past, as does buying the last stamp needed to complete a collection. Putting aside the past removes events on which these achievements depend and so puts aside the achievements. A satisfactory method of skipping the past uses realizations of basic intrinsic attitudes to identify considerations to process and to put aside. It uses the past to identify future realizations of basic intrinsic attitudes and then puts aside past realizations of basic intrinsic attitudes. Basic intrinsic attitudes set the stage for putting aside past events in an effective and justified way.

Expected-utility analysis, as Chapter 2 presents it, takes an option's possible outcomes to be possible worlds. Intrinsic-utility analysis also evaluates an option's world. A world's evaluation may put aside the world's realizations of extrinsic desires and aversions. The world's evaluation depends just on its realizations of basic intrinsic attitudes. For example, if a world realizes both an agent's extrinsic desire for money as a means to pleasure and the agent's intrinsic desire for pleasure, an evaluation of the world should not count realization of the extrinsic desire in addition to counting realization of the intrinsic desire. Doing so double counts the intrinsic desire because it grounds the extrinsic desire.

Although neither expected-utility analysis nor intrinsic-utility analysis narrows the scope of an option's evaluation, they form the foundation for more focused types of utility analysis. Three types of utility, having evaluative scope narrower than the scope of ordinary, comprehensive utility, facilitate utility comparisons of options. The first type of utility takes a temporal perspective, the second a spatiotemporal perspective, and the third a causal perspective. Comprehensive utility evaluates an option by considering the possible world that would be realized if the option were realized, the option's world. The new types of utility consider, not the option's world, but a temporal, spatiotemporal, or causal segment of that world. Each successive type further reduces the scope of an option's evaluation. Each comes with a method of utility analysis for dividing a utility of its type. The results are temporal-, spatiotemporal-, and causal-utility analyses. Expected-utility analysis and intrinsic-utility analysis justify these specialized types of utility analysis.

Common methods of evaluating an option treat the option's sequel, events it may influence, or events that are its consequences. Chapters 4–6 explicate these methods using specialized types of utility. They show how deliberations about decisions commonly use temporal-, spatiotemporal-, and causal-utility analyses to simplify evaluations of options. The chapters on these forms of utility analysis explicate and justify these methods of separating considerations. They introduce the forms of analysis first for cases in which an agent is certain of each option's world and then for cases with uncertainty about some option's world. They handle the latter cases by combining the specialized forms of utility analysis with expected-utility analysis.

The hedonic calculus divides the utility of a period into the utilities of subperiods, with experiences during a period or subperiod yielding its utility. Chapter 4 presents temporal-utility analysis, which the remarks on separation briefly described. This type of analysis, a descendant of the hedonic calculus, separates events according to times of occurrence. Consider planning a garden. The plan's formation takes place during a period of time. Events at the beginning of the period, scattered thoughts, have little value. The plan's completion has value, however. A temporal evaluation of the period considers the value of events in the period. Applying a temporal analysis to a world, its utility is the sum of the temporal utility of past events and the temporal utility of other events. An evaluation of an option may drop the temporal utility of past events. Done for all options, this amounts to a scale change for comprehensive utility. Focusing on the future does not change the options' ranking.

Although people sometimes evaluate an option by considering the course of events in a spatiotemporal region following the option's realization, the literature does not have a detailed treatment of this method of utility analysis. Chapter 5 presents spatiotemporal-utility analysis, which divides an option's results according to the spatiotemporal regions in which they occur. Consider planting a garden. The work takes place during a period of time and in a volume of space. Planting the beans may have more value than planting the radishes. A spatiotemporal evaluation of a spatiotemporal region considers the value of events in the region. Applying a spatiotemporal-utility analysis to an option's world, its utility is the sum of the spatiotemporal utility of events in the option's region of influence (its future-directed light cone according to relativity theory) and the spatiotemporal utility of events outside its region of influence. An option's evaluation may drop the spatiotemporal utility of events outside the option's region of influence. Done for all options, this amounts to a scale change for comprehensive utility and does not change the options' ranking. Our deliberations often use spatiotemporal separation of considerations because we know that it is harder to affect distant events than near events. Space and time limit our reach, we know, even if we do not know precisely our reach's spatiotemporal limits.

Chapter 6 treats causal-utility analysis. Consider weeding the garden. The effects include more nutrients for the beans and radishes but exclude the rain that follows the weeding. The rain is not a consequence of the weeding. An efficient evaluation of weeding focuses on weeding's consequences. Applied to a world, causal-utility analysis separates the utility of an option's world into the causal utility of the option's consequences and the causal utility of other events. An option's evaluation may drop the causal utility of other events. Done for all options, this amounts to a change in scale for comprehensive utility and does not change the options' ranking.

Causal decision theory inspires causal-utility analysis. Its version of expected-utility analysis claims that an option's consequences settle utility comparisons with other options. A principle of evaluation for the options in a decision problem may therefore focus on options' consequences and trim all other considerations, including the past, that no option influences. The evaluation requires separating the events that would occur if an option were realized into the option's consequences and other events.

Gibbard and Harper ([1978] 1981: 167), in their famous paper on causal decision theory, suggest separating into parts an option's utility using the distinction between an option's consequences and its unavoidable outcomes. Chapter 6 elaborates and refines their suggestion. Causal-utility analysis separates an option's utility into the causal utility of its consequences and the causal utility of its unavoidable outcomes. The chapter's illustrations assume that an option's unavoidable outcomes include the past, events outside the option's light cone, and events in an option's light cone that the agent cannot alter by realizing a different option. Even if the chapter's assumptions about unavoidable outcomes do not hold in worlds with backward causation and causation faster than light, the method of excluding unavoidable outcomes works in those worlds. The chapter's assumptions about unavoidable outcomes merely ground illustrations of causal-utility analysis.

Causal-utility analysis rests on intrinsic-utility analysis. It uses causal relations to identify a minimum set of basic intrinsic attitudes such that the intrinsic utilities of their realizations and the probabilities of their realizations yield an option's utility on a scale suitable for comparing options to identify rational options. Because many of an option's consequences are matters of indifference, an option's evaluation may consider just basic intrinsic attitudes whose realizations may be the option's consequences. The combination of causal- and expected-utility analyses is the book's ultimate simplification of an option's evaluation. It is the most efficient general type of evaluation because it uses causal utility, whose scope is the minimum sufficient for a reliable ranking of options in all decision problems. Although a comparison of two options may ignore shared consequences, an option's evaluation may not ignore any of its consequences without jeopardizing the evaluation's accuracy.

Less efficient simplifications, considered along the way to the combination of causal-utility analysis and expected-utility analysis, still deserve a place in decision theory's toolbox. Temporal and spatiotemporal methods of analysis, although not optimally efficient, are easier to apply than are causal methods because they do not require identifying an option's consequences. Sometimes an agent may be in a position to apply temporal- or spatiotemporal-utility analysis, but not causal-utility analysis, because the agent can identify the past or options' regions of influence but cannot identify options' consequences. Evaluations that use options' futures or regions of influence, even though they are not maximally efficient, are more efficient than world-Bayesianism and sometimes are easier to apply than causal-utility analysis. An agent may not have enough information to apply the simplest general form of utility analysis. Having multiple methods of simplifying an option's evaluation makes decision theory versatile. It may use the method that best fits a particular decision problem.

Also, gains in efficiency for evaluation of options come by paying a conceptual price. Temporal-, spatiotemporal-, and causal-utility analyses carry, respectively, conceptual commitments to temporal, spatiotemporal, and causal relations. Maximum efficiency in evaluation of options has the highest conceptual price. Temporal- and spatiotemporal-utility analyses make weaker metaphysical assumptions than does a causal-utility analysis. Time's existence, for example, has a firmer metaphysical foundation than does causation's existence. Metaphysical scruples may recommend methods of evaluating an option that sacrifice efficiency to lower conceptual costs. These methods may be maximally efficient given certain metaphysical constraints. For example, spatiotemporal-utility analysis yields a maximally efficient evaluation putting aside evaluations that use causality.

Furthermore, a person often evaluates options using their futures and regions of influence, and an account of these evaluations has intrinsic interest even if the evaluations are not maximally efficient. Temporal- and spatiotemporal-utility analyses are worth studying independently of their role as stepping-stones to causal-utility analysis's maximally efficient method of evaluating options.

Special features

Intrinsic-utility analysis breaks down a possible world's utility according to the basic intrinsic desires and aversions that the world realizes, and expected-utility analysis breaks down an option's utility according to its possible outcomes. The combination of these forms of utility analysis uses both the dimension of basic intrinsic attitudes and the dimension of possible outcomes. It is *multidimensional.*

Multi-attribute–utility analysis divides a world's utility into the utilities of the world's attributes. Intrinsic-utility analysis does the same using realizations of basic intrinsic attitudes as the relevant attributes. The attributes form a single dimension of utility analysis. In contrast with multi-attribute–utility analysis, multidimensional-utility analysis locates utilities along multiple dimensions, such as attribute and modal dimensions. The dimension of attributes is just one dimension among many along which an analysis may spread a world's utility.

Temporal-, spatiotemporal-, and causal-utility analyses spread utilities across time, space-time, and consequences and thereby introduce temporal, spatiotemporal, and causal dimensions of utility analysis to supplement the dimensions of possibilities and realizations of basic intrinsic attitudes. The additional dimensions provide new locations for utility but not new sources of utility. All the reasons behind an option's evaluation that the new forms of utility analysis entertain reduce to reasons concerning probabilities of realizing basic intrinsic attitudes. For instance, temporal-utility analysis presumes an assignment of a basic intrinsic attitude's realization to a time and then treats the time rather than the attitude. The time to which the analysis assigns a utility is a location but not a source of utility. It is not even a component of a reason for a proposition's utility (except in special cases in which basic intrinsic attitudes target times). Its utility depends on the utilities of possible outcomes and realizations of basic intrinsic attitudes. The utilities that temporal-, spatiotemporal-, and causal-utility analyses spread along dimensions derive from the utilities of chances for realizations of basic intrinsic attitudes.

The various ways of separating an option's utility into parts combine with one another. Temporal-utility analysis combines with expected-utility analysis, for example. For each type of utility analysis, no matter whether it uses comprehensive, intrinsic, temporal, spatiotemporal, or causal utility, its combination with expected-utility analysis handles cases with uncertainty about an option's world.

The various methods of analyzing options' utilities are consistent. The chapters' appendices show that the methods yield consistent comparisons of options by showing that the methods all follow from the same combination of intrinsic-utility analysis and expected-utility analysis. The appendices do not explicitly treat all combinations of utility analyses, but their methods of proof extend to all combinations.

This introduction presents the problem the book treats, namely, efficient evaluation of options, and the book's method of solving that problem, namely, separation of considerations. The following chapters present the solution in detail. They introduce types of utility of narrow scope and types of utility analysis that break the utility of a whole into the utilities of its parts. Progress toward efficiency begins with general principles for separating considerations for and against an option, principles that justify ignoring irrelevant considerations. The next chapter formulates these principles.

CHAPTER I

Separability

Efficiency simplifies an option's evaluation by trimming considerations that affect all options equally. A preliminary step separates considerations into those to be processed and those to be put aside. An accurate evaluation processes considerations that are separable or independent from those dropped.

Imagine composites divided into parts of specific types, for example, baskets of groceries divided into cereal, dairy, meat, and produce parts. Given a part's separability from other parts, a ranking of composites according to the part's contents agrees with a ranking of composites in which the other parts have constant contents. For example, if the cereal part is separable from the other parts, then ranking baskets according to contents of the cereal part agrees with ranking baskets that have constant contents in the other parts. Focusing on the cereal part is a shortcut to ranking baskets.

This chapter treats separability of considerations generally, exploring its philosophical foundations rather than its technical applications. It treats separability generally so that its applications to probability and various types of utility have behind them the explanatory power of general principles. Separability belongs to a cluster of similar phenomena, and exploring its relation to the cluster's other elements clarifies it. After defining a type of separability, the chapter treats separability's relation to complementarity and compositionality. It presents the conceptual geography of separability to create a context for arguments supporting separability and to identify promising argumentative strategies. The chapter finishes with a review of methods of establishing separability. It shows how to establish separability as a norm for ideal agents. Later chapters construct arguments for specialized norms of separability, grounding, for example, expected-utility analyses of comprehensive utilities. They introduce various types of utility and argue for the principles of separability that apply to them.

In deliberations, separating considerations is the first step toward efficiency; thus, this chapter introduces the technical concept of separability at length. Later chapters use principles of separability to support proposals about efficient evaluation of the options in a decision problem.

1.1 Separability's definition

Assuming complete information, an option's comprehensive utility evaluates the option's world. A narrower evaluation targets the option's future instead of the option's world. It guides decisions well if the order of options according to futures is the same as the order of options according to worlds. In general, simplified deliberations may evaluate an option using part of the option's world instead of the option's entire world if evaluations using the part replicate comprehensive utility's order of options using worlds.

This method of simplifying choices assumes that a part's evaluation affects the world's utility independently of the other parts. The type of independence that justifies putting aside considerations is called *separability*. Works in economics on separability and in measurement theory on decomposability, another name for separability, generally define utility using preferences. As Chapter 2 explains, a realist interpretation of utility that defines utility using degrees of desire rather than preferences strengthens norms of utility. This section modifies the literature's definition of separability to suit a definition of utility using degrees of desire. Also, the literature adjusts separability's definition according to separability's role in a theory. This section's definition tailors separability for its role in simplifying deliberations.[1]

The chapter's main target is separability of utility. The definition of separability, being general, applies to all types of utility, including utility attaching to goods as well as utility attaching to propositions, although later chapters use exclusively utility attaching to propositions, taking propositions to represent worlds, outcomes, and possession of goods.

The definition of utility's separability derives from the separability of parts of composites in a preference ranking. For uniformity among the objects of preference and utility, preferences hold between realizations of propositions when utility attaches to a proposition's realization. A preference for the truth of one proposition rather than another shows in decisions to realize the first proposition rather than the second and typically arises from beliefs about the consequences of each proposition's truth. A norm requires preferring one proposition's realization to another's realization if

---

[1] Topology uses another kind of separability, namely, a property of metric spaces limiting their size, to define separable spaces.

the desire for the first's realization is stronger than the desire for the second's realization. Chapter 2 elaborates these points.

Although this section's objective is to provide an account of utility's separability, it starts with an account of separable preferences among composites. To obtain this account, it formulates a general definition of separability for an order of composites.

1.1.1   Separable orders

Reviewing cases of separability and the general features of separability reveals two candidates for separability's definition. One defines separability as an order's independence from conditions. The other defines separability as a relation between an order and a suborder.

Suppose that wholes and their parts have various instantiations, as a two-part commodity bundle of apples and bananas has various instantiations depending on its number of apples and number of bananas. Also suppose that the instantiations of wholes have an order, that for each part its instantiations have an order, and that for each subset of parts its instantiations have an order. Preferences may supply the order. Separability of parts and subsets of parts holds with respect to an order of instantiations of the wholes, their parts, and subsets of their parts. For example, it may apply to parts of commodity bundles as preferences order their instantiations. Agreement of the order of a part's instantiations with the order of compatible instantiations of wholes, for all instantiations of the other parts, signals the separability of the part from the other parts. In two-part commodity bundles of apples and bananas, apples are separable from bananas if the order of the bundles goes by the number of apples, for any constant number of bananas. A set of parts, not just a single part, is separable from a whole's other parts just in case the order of instantiations of the set's parts agrees with the order of compatible instantiations of wholes, for any fixed instantiation of the other parts.[2]

A set of n-tuples of values of variables, or n-placed vectors with occurrences at locations in the vector, may represent a set of composites. The

---

[2] McClennen (1990) defines separability for preferences at different moments of time. His page 12 says that given separability, preferences at a future time depend only on consequences realizable at the future time and so are independent of earlier preferences. His pages 120–21 say that separability holds if and only if choice does not depend on earlier choices. His page 122 characterizes separability using decision trees: past choices reduce a tree without changing the tree's remainder. Hammond (1988) and Seidenfeld (1988) define similar types of temporal separability for preferences. Hammond calls his type of temporal separability *dynamic consistency*, and Seidenfeld calls his type of temporal separability *dynamic feasibility*.

Table 1.1 *Illustration of terminology*

(x, y, z)	vector of variables, a composite's structure
(x_1, y_1, z_1)	vector of variables' values, a composite
(x, y)	subvector of variables, a subset of parts' structure
(x_1, y_1)	subvector of variables' values, a subset of parts' instantiation

sequence of variables may represent the sequence of a composite's parts, if they come in a sequence. For example, a vector giving the amount of each good in a set of n goods represents a shopping basket of the goods. A variable represents each good, and a value of the variable stands for the amount of the good in the basket. The vector's sequence of goods represents their sequence, if any, in the basket. A consumer's preferences order possible shopping baskets and so order vectors that represent the baskets.

Table 1.1 illustrates this terminology. The variables x, y, and z have as values, respectively, x_1, y_1, and z_1. A composite's structure has parts with variable values. Variables represent the parts, and vectors of variables represent the structure of composites. Assigning values to the variables produces a vector of values that represents a composite. A subvector of variables represents a subset of parts' structure, and a subset of values of variables represents an instantiation of the parts.

Given an order of vectors of values of variables and an order of subvectors of values of subsets of the variables, a subset of variables is separable from the other variables if and only if, for all values of the other variables, the order of subvectors of values of the subset of variables agrees with the order of vectors of values of all variables. Hence, for a vector of locations, a subvector of the locations, and corresponding vectors and subvectors of occupants of the locations, the subvector of locations is separable from its complement if and only if the order of subvectors of occupants agrees with the order of vectors of occupants. For example, suppose that the variables x and y represent two goods. The constants x_1 and x_2 represent values of x, and the constants y_1 and y_2 represent values of y. The vectors of values of the variables are (x_1, y_1), (x_2, y_1), (x_1, y_2), (x_2, y_2). Suppose that the list gives their order from lowest to highest so that by using the symbol "<" to represent their order, it would be $(x_1, y_1) < (x_2, y_1) < (x_1, y_2) < (x_2, y_2)$. For any value of y, the order of x's values given by their subscripts agrees with the order of vectors. Hence, x is separable from y. Taking the subscripts of y's values to show their order, y is also separable from x because given each value of x, the ranking of y's values agrees with the ranking of pairs. Separability exists within a set of variables generating the components of composites, given a way of ordering the composites and subsets of their components.[3]

The agreement of orders that separability entails may hold by definition or as a consequence of separability's definition and features of the orders. Rather than define separability as an agreement of orders, some theorists define it as an independence of conditional orders that they define using a nonconditional order.

The order of vectors of values of all variables defines an order of subvectors of values of a subset of variables given a way of fixing the subvectors' complement. That is, it defines a *conditional order* of the subvectors. In the example, $\{x\}$ is a subset of variables, and (x_1) and (x_2) are subvectors of its values. The definition declares that (x_2) ranks above (x_1) when y's value is y_1 because vectors of values of x and y containing x_2 rank above vectors of values of x and y containing x_1 when y's value is y_1. In the example, for all ways of fixing the value of y, the vectors of values of x and y containing x_1 have the same rank with respect to vectors of values of x and y containing x_2. If the conditional order of subvectors defined using the order of vectors is the same for all values of y, then the common conditional order of subvectors may define the subvectors' nonconditional order.[4] The subvectors' order so defined agrees with the order of vectors; the variable x is separable from the variable y. The variable x's separability from the variable y amounts to the vectors' order generating the same order of subvectors of x's values given any value of y. Starting with the order $(x_1, y_1) < (x_2, y_1) < (x_1, y_2) < (x_2, y_2)$, the conditional order $(x_1) < (x_2)$ given y_1 derives from the order of pairs. Also, the conditional order $(x_1) < (x_2)$ given y_2 derives from the order of pairs. The variable x is separable from the variable y because these conditional orders of (x_1) and (x_2) agree. The variable y is similarly separable from the variable x. In general, the order of subvectors of values of a subset of variables is the same given all ways of fixing the values of the other variables — that is, the conditional orders of the subvectors are the same given all ways of fixing the values of the other variables — just in case the variables generating the subvectors are separable from the other variables.

To illustrate, imagine a diner who prefers a ham sandwich to a beef sandwich, whether or not the sandwiches have cheese, but prefers each type

---

[3] Broome (1991: 22–25, 65–80) states that in a set of n-placed vectors, a set of subvectors is separable if and only if comparison of the subvectors, conditional on a constant complement of the subvectors, is the same as comparison of the whole vectors. Imagine an order of a set of vectors. A subset of locations in these vectors is separable if and only if the order of subvectors of occurrences at these locations, given all ways of fixing occurrences at other locations, agrees with the order of vectors.

[4] Broome (1991: 67) defines the order of subvectors using the order of vectors.

of sandwich with cheese. This order of vectors represents the diner's prefer-ences: (no cheese, beef) < (cheese, beef) < (no cheese, ham) < (cheese, ham). Applying the definition of conditional preference, using these nonconditional preferences, the diner prefers cheese to no cheese given beef and prefers cheese to no cheese given ham. Because the conditional preference for cheese holds regardless of the meat, the variable indicating whether the sandwich includes cheese is separable from the variable indicating the type of meat.

The order of vectors need not define the order of subvectors, however. For example, the diner may prefer cheese to no cheese even if the cheese is not in a sandwich. The preference for cheese need not derive from the preference ranking of sandwiches.[5]

If the order of vectors does not define the order of subvectors, then conditional independence of the order of subvectors does not entail its agreement with the order of vectors. Conditional independence is not sufficient for agreement. The order of subvectors may be constant for any way of fixing the subvectors' complement but contrary to the order of vectors. In the example, the conditional independence of the order of x_1 and x_2 with respect to the value of y authorizes using the order of pairs to define the order $x_1 < x_2$. However, the order derived from the order of pairs may disagree with the actual order $x_1 > x_2$. For example, taken alone, less pepper may taste better than more pepper, although taken with either a carrot salad or a preferred beet salad, two dashes of pepper may taste better than one dash of pepper. A set of vectors creates a context for their elements' realizations, and the context may influence preferences among their elements' realizations.[6]

A definition of separability that takes it as conditional independence of an order of subvectors does not require agreement of orders of subvectors and vectors unless the order of vectors defines the order of subvectors. Because agreement of orders, rather than conditional independence, is crucial for simplifying deliberations, to obtain it from separability without using the order of vectors to define the order of subvectors, this section defines

---

[5] Preference may also order subvectors that a subset of variables generates given a set of values for the other variables. Preference's order may differ from the order of subvectors that the order of vectors imposes given the values of the other variables, although for a rational ideal agent the two orders of subvectors agree.

[6] Binmore (2009: 47–49) takes separability for an order of pairs to obtain if preferences concerning one factor are independent of the value of the other factor. The pairs may involve goods or gambles concerning goods. For a set of pairs, let the relation < represent preference for the pair on the right, and the relation ≤ represent preference for the pair on the right or indifference between the pair on the left and the pair on the right. If two variables with L and L' and M and M' as values, respectively, are separable from each other, then $(L, M) < (L, M')$ implies $(L', M) \le (L', M')$, and $(L, M) < (L', M)$ implies $(L, M') \le (L', M')$.

separability as agreement of orders and not conditional independence. It is then a relation between an order of subvectors and the order of vectors, rather than a feature of the order of vectors. Accordingly, separability has this definition:

> In a set of variables, a subset is *separable* from the others if and only if for all ways of fixing the values of the other variables, the order of subvectors of values of variables in the subset agrees with the order of vectors of values of all variables.

In the example about the variables x and y, the order of vectors of their values is $(x_1, y_1) < (x_2, y_1) < (x_1, y_2) < (x_2, y_2)$. Suppose that each variable's values are amounts of an economic good and that preferences order its amounts taking the good in isolation. A consumer prefers more to less of good x and also of good y, and the subscripts of a variable's values indicate amounts of the good the variable represents, so that the values of the variables have the order of their subscripts. Then, x is separable from y because the order of subvectors of x's values agrees with the order of vectors of values of all variables for any fixed value of y.[7]

Separability taken as a relation between an order of vectors and an order of subvectors is equivalent to separability taken as a property of the order of vectors, if the order of vectors defines a conditional order of subvectors that in turn defines the nonconditional order of subvectors. According to separability taken as a property, in a list of variables a subset of variables is separable from the others if and only if no matter how the others are fixed, the subvectors of values of the variables in the subset have the same order. The subvectors' constant conditional order defines their nonconditional order. Their nonconditional order agrees with the order of the vectors because the order of vectors defines the order of the subvectors of values of variables in the subset conditional on fixed values of the other variables. Hence, given the definitions, the subvectors of values of the variables in the subset have the same order no matter how the other variables are fixed if and only if they have a nonconditional order that agrees with the order of the vectors.

---

[7] For a statement of the definition in technical notation, consider a vector X of variables $X_1, X_2, \ldots X_m$, a subvector $\mathbf{X}_i$ of these variables, and the subvector's complement $\mathbf{X}_j$ with respect to X. x_i and $x_{i'}$ are values of a variable X_i belonging to $\mathbf{X}_i$, and x_j is a value of a variable X_j belonging to $\mathbf{X}_j$. $\mathbf{x}_i$ and $\mathbf{x}_{i'}$ are respectively vectors of values x_i of variables in $\mathbf{X}_i$ according to the variables' order in $\mathbf{X}_i$. $\mathbf{x}_j$ is a vector of values of variables in $\mathbf{X}_j$ according to the variables' order in $\mathbf{X}_j$. $\mathbf{x}_{ij}$ is the vector of values of the variables in X according to the variables' order in X formed from the subvectors $\mathbf{x}_i$ and $\mathbf{x}_j$, and $\mathbf{x}_{i'j}$ is a similar vector formed from the subvectors $\mathbf{x}_{i'}$ and $\mathbf{x}_j$. Let $\geq$ represent the order of vectors and subvectors of values. The subvector of variables $\mathbf{X}_i$ is *separable* from its complement $\mathbf{X}_j$ if and only if for any subvector $\mathbf{x}_j$, for all subvectors $\mathbf{x}_i$ and $\mathbf{x}_{i'}$, $\mathbf{x}_i \geq \mathbf{x}_{i'}$ if and only if $\mathbf{x}_{ij} \geq \mathbf{x}_{i'j}$.

To illustrate the equivalence, consider preferences among worlds taken as conjunctions of a past and a future (that includes the present). Values of a pair of variables – p for the past and f for the future – form a vector that represents a world. Suppose, using < between two propositions to represent preference for the second's truth to the first's truth, that the conditional preference $f_1 < f_2$ given p_1 by definition derives from the preference between conjunctions $(p_1 \ \& \ f_1) < (p_1 \ \& \ f_2)$, and similarly for other conditional preferences. By definition, the order of worlds settles the order of futures given the past. Also, suppose that the future is separable from the past in the property sense. Given this separability, $f_1 < f_2$ given p_1 if and only if $f_1 < f_2$ given p_2 or, equivalently, $(p_1 \ \& \ f_1) < (p_1 \ \& \ f_2)$ if and only if $(p_2 \ \& \ f_1) < (p_2 \ \& \ f_2)$. The order of futures conditional on the past, as derived from the order of worlds, is the same for all ways of fixing the past. Take the nonconditional order of futures as that constant conditional order. Then, the nonconditional ranking of futures agrees with the ranking of worlds whatever the past. So, the future is separable from the past in the relational sense. Similar inferences move from separability in the relational sense to separability in the property sense.

In a decision problem, when using evaluation of options' futures to simplify evaluation of options, it is best not to use the order of options' worlds to define the order of options' futures. Applying the definition to obtain the order of futures requires obtaining the order of worlds, and so it does not simplify evaluations. To simplify evaluations, it is best to infer the order of worlds from a definitionally independent order of futures, using the future's separability from the past in a relational sense. In general, a world part's separability from its complement in the relational sense justifies evaluations that attend only to the part.

Separability is a type of independence for components of composites. Elegance suggests using the order of composites to define conditional orders of composites, independence of these conditional orders, and then separability of components. However, the project of simplifying choices needs a nonderivative type of independence of components. Because simplification derives the order of composites from the order of components, it requires that the order of components be definitionally independent of the order of composites. Hence, it requires the relational definition of separability.

1.1.2 Utility

An order is separable just in case it orders composites that have parts that are separable with respect to the composites' order and the parts' order. Separability of a part entails independence of the order of its instantiations from other parts' instantiations. This independence entails an analogous independence of a function representing the order of the part's instantiations, in particular, a utility function representing preferences among the part's instantiations. Separability of a set of parts also entails an analogous independence of functional representations of the order of the set's instantiations.

Suppose that a commodity bundle's utility is a sum of its parts' utilities. Then, each set of parts is separable from its complement, using the preferences that a utility function represents to order instantiations of wholes and sets of parts. Ordering a part's instantiations according to preferences agrees with ordering particular bundles containing its instantiations according to preferences, given that the other parts have a constant instantiation. For example, if x_2 has greater utility than x_1, then, no matter what the utility of the variable y's value, the pair (x_2, y) has greater utility than the pair (x_1, y) because by supposition the utility of a pair equals the sum of its constituents' utilities.

Given separability of a world's part from the world's other parts, the order of the part's instantiations agrees with the order of worlds for all instantiations of the other parts. Given the future's separability from the past, the order of futures agrees with the order of worlds for every account of the past. A world's parts are separable if evaluations of a world's parts sum to the world's utility. The world's utility increases as a part's evaluation increases, given any way of fixing other parts. So, the order of the part's instantiations, holding other parts fixed, agrees with the order of worlds containing the part's instantiations.

Separability of an order of vectors constrains utility functions representing the order of vectors. So, separability may extend from orders of vectors to utility functions representing the orders. This section introduces separability of an ordered subset of the argument variables of a utility function representing an order of vectors. It assumes that a utility function exists over subvectors of values of the subset of argument variables that represents the order of the subvectors, but it does not assume that the order of the vectors and their utilities defines the order and the utilities of the subvectors. A proposition may express the state any vector or subvector represents, so utility defined using strength of desire applies to the vectors and subvectors through the propositions that represent them.

Separability of variables giving the arguments of a utility function follows from separability of variables generating the vectors to which the utility function attaches and whose order the utility function represents. Take the function $U(x, y)$. The variable x is separable from the variable y just in case x

is separable from y with respect to the vectors (x, y) whose order U represents. For a utility function, separability of a subset of argument variables from its complement holds if and only if the order of subvectors the variables generate agrees with the order of vectors no matter how the other variables are fixed. Given separability, moving up the order of subvectors, holding their complements constant, entails moving up the order of vectors, so an equivalent characterization of the subset's separability uses *strictly increasing monotonicity*: increasing the utility of the subset of argument variables, holding fixed argument variables in its complement, increases the function's value. If the function is strictly monotonically increasing at an argument place, increasing the argument's value increases the function's value, holding fixed the values of arguments at other places. The monotonicity condition entails the separability of the argument place, and because its separability entails the monotonicity condition, the monotonicity condition is equivalent to the argument place's separability. A similar equivalence holds for a set of argument places.

Suppose that the utility of composites is a *separable function* of the utilities of components. Then, the function is strictly monotonically increasing at each argument place. This holds if the function is addition. Utility's additivity, if it obtains, implies the utility function's separability.

When a subset of variables giving arguments of a utility function is separable from the other variables, some function of the utilities of the subvectors and of their complements yields the utilities of the vectors. For example, $U(x, y) = F(U(x), y)$ for some F given x's separability from y. Given that separability, a utility subfunction $U(x)$ represents the order of x's values, that is, the subvectors (x), given any value for y. The utility subfunction replaces the variable x in the move from $U(x, y)$ to $F(U(x), y)$. The function $F(U(x), y)$ represents the order of vectors, and, with respect to that function, $U(x)$ is separable from y. Given a fixed value of y, if $U(x)$ increases, then so do $F(U(x), y)$ and $U(x, y)$. A subset of a utility function's argument variables is separable from the others if and only if the corresponding function of the utilities of the subsets' values and of their complements is strictly monotonically increasing in the utilities of the subsets' values given any way of fixing its complement's values. Separability of utility functions of variables and of utility functions in which a subutility function replaces a set of variables arises from the separability of the variables.[8]

---

[8] Varian (1984: sec. 3.14) defines utility's *functional separability* as the utility of a composite's being increasing given increases in a component's utility. The utilities of components are functionally separable if and only if holding constant all components but one, while increasing the utility of that

Given that utility settles order, in the order of vectors (x, y) the variables x and y are mutually separable if and only if in the utility function $U(x, y)$ the variables are also mutually separable. Also, the arguments of the utility function $U(x, y)$ are mutually separable if and only if the arguments of the function $F(U(x), U(y))$ are mutually separable. Hence, $U(x)$ and $U(y)$ are mutually separable in $F(U(x), U(y))$ if and only if x and y are mutually separable in (x, y). This equivalence holds for n-tuples as well as pairs.

For a preference order of vectors of two variables, this section distinguishes one variable's effect on preferences concerning the other variable from its effect on utilities concerning the other variable because it takes utilities to represent not just preferences but also strengths of desire. According to this section's accounts of separability and utility, for mutually separable variables x and y the equation $U(x, y) = F(U(x), U(y))$ does not state a feature of the order of vectors of values of x and y, but instead it states a relation of three utility functions not defined by the order of vectors. An argument for the equation appeals not just to features of the order of vectors but also to relations between utility functions that represent the order of vectors and the orders of subvectors.

The literature on separability often treats utility attaching to combinations of goods, but points about separability carry over to utility attaching to conjunctions of propositions. Many structural points about utility apply whether utility attaches to goods and combinations of goods or to propositions and conjunctions of propositions. Consider instead of $U(x, y)$, with utility applying to pairs of goods, $U(x \,\&\, y)$, with utility applying to conjunctions of propositions.

If a type of utility U applies to binary conjunctions, the value of the variable x yields the first conjunct, the value of the variable y yields the second conjunct, and the variables are mutually separable, then $U(x_1 \,\&\, y_1) = F(U(x_1), U(y_1))$ and $U(x_1 \,\&\, y_2) = F(U(x_1), U(y_2))$ for some strictly monotonically increasing function F. To compare $U(x_1 \,\&\, y_1)$ with $U(x_1 \,\&\, y_2)$, letting F be the reduction of F to a one-place function given $U(x_1)$ as fixed first argument of F, one may compare $F^*(U(y_1))$ with $F^*(U(y_2))$ or equivalently $U(y_1)$ with $U(y_2)$ because F is strictly monotonically increasing. By separability, the order of conjunctions according to F agrees with the order of second conjuncts according to $U(y)$ after fixing the value of x.

A simplified evaluation of composites using separability excises the common element from the composites and compares them using the

component, a composite's utility increases. According to utilitarianism, individuals' utilities are functionally separable. The utility function for one individual settles the ranking of worlds given fixed utilities for other individuals.

remainders' utilities, according to an appropriate type of utility. Chapter 4 compares worlds by excising the past from the worlds and evaluating their futures. Suppose that p stands for the past and f stands for the future. Given the mutual separability of f and p, because $U(p, f) = F(U(p), U(f))$, the order of (p_1, f_1) and (p_1, f_2) follows from the order of f_1 and f_2. Moreover, because of separability, if $F(U(p_1), U(f_1)) \leq F(U(p_1), U(f_2))$, then $F(U(p_2), U(f_1)) \leq F(U(p_2), U(f_2))$. The utilities of futures settles the utilities of worlds no matter what the past is like. Hence, in a decision problem, deliberators may ignore the past given the utilities of futures.

Separability in this chapter's sense imposes constraints on utility functions for vectors and subvectors; however, it does not entail that the utility curve for a subvector has the same shape for all ways of fixing its complement. For vectors (x, y), the shape of $U(x)$ may depend on the value of y. Then, the utility of a value of x depends on the context, even though the values of x have a constant order. Consider baskets of apples and bananas. A basket's utility depends on the basket's number of apples and number of bananas. Fix the number of bananas and so the utility of bananas. Then, a basket's utility increases as the number of apples increases, but the increase in its utility depends on the number of bananas in the basket. The greater the number of bananas in the basket, the less utility more apples add to the basket. The number of bananas affects the utility of apples. So, the shape of the utility curve for apples changes as the number of bananas in the basket changes. An apple's marginal utility is not independent of the number of bananas. It declines as the number of bananas increases. The utility curve for apples flattens as the number of bananas increases. However, for each quantity of bananas, the order of baskets agrees with the order of quantities of apples. Increasing the quantity of bananas, the utility of apples goes down, but the preference order is the same – the more apples the better.

The effect carries over to cases in which probabilities govern possible outcomes. In such cases, a *gamble* in the economic sense represents a prospect of gaining apples by specifying the probability of each number of apples that may issue from the prospect. The quantity of bananas may affect preferences among gambles concerning apples, even though the agent still prefers more apples to fewer so that the type of separability this chapter introduces obtains.[9]

---

[9] Some definitions of separability attend to marginal utility. Black (2002) states that if x and y are variables having as values amounts of a commodity, a utility function for a pair (x, y) is separable if and only if the marginal utility of x is independent of y, and the marginal utility of y is independent of x. Gollier (2001: 202) states that apples and bananas are not separable if the marginal utility of apples

Let us call the type of separability that preserves order *ordinal separability* and a more demanding type that preserves the shape of the utility function *metric separability*. Metric separability assumes that a utility function represents more than order. It may represent intensities of preference as well as preferences. Ordinal, but not metric, separability holds for the baskets of apples and bananas. Because ordinal separability suffices for simplification of choices, I treat only it.

1.1.3   Complementarity

Some types of *complementarity* oppose separability. Suppose that a first variable has either a left glove or a right glove as value, and similarly a second variable has either a left glove or a right glove as value. The order of pairs formed using a value of each variable does not generate a single order of the first variable's values for all ways of fixing the second variable's value. The first glove is not separable from the second glove because the order of values for the first glove, right and left, depends on whether the value of the second glove is right or left. Suppose that the order of pairs is (l, l), (r, r), (l, r), (r, l) from lowest to highest, except with indifference between the last two pairs. With a left second glove, the order puts a right first glove higher than a left first glove. In contrast, with a right second glove, it puts a left first glove higher than a right first glove. No way of defining the order of first elements and the order of second elements makes the order of either set of elements agree with the order of pairs given each element of the other set. Neither variable is separable from the other because the values of the two variables are complementary; the good pairs have one right glove and one left glove.[10]

Not all types of complementarity oppose separability. Habit creates complementarity between past and future events. Because of habit, past

varies with the number of bananas already in a consumption bundle so that the value of an extra apple depends on how many bananas are in the bundle. Blackorby, Primont, and Russell (2008) define separability in terms of marginal rates of substitution of commodities. Two variables are separable from a third if and only if the marginal rate of substitution between the two variables is independent of the third. This type of separability implies an aggregation function (a subutility function) over the two variables that is independent of the third. For a single variable, its separability from the others is its having the same marginal utility given fixed values of the other variables. The definition assumes that the values of variables are continuous so that marginal rates of substitution are defined. Separability defined in terms of marginal utilities puts subutility functions for subvectors on the same scale and grounds a representation of an order of vectors that adds the subutilities of subvectors partitioning a vector to obtain the vector's utility.

[10] The order of pairs of gloves need not define the order of gloves in isolation. Preferences may order gloves taken in isolation and not just pairs of gloves. In a rational ideal agent, the order of single gloves may derive from each glove's evaluation according to a probability-weighted average of evaluations of pairs it may form.

activities affect the value of future activities. A typical person enjoys exercising an acquired skill. For example, a tennis pro enjoys playing tennis. Past training affects the value of future tennis. Despite this complementarity between past and future events, the order of worlds agrees with the order of futures. When future tennis increases the value of a future, it also increases the value of a world with the future given any compatible settling of the past. This correlation of values makes the future separable from the past. Chapter 4 elaborates the point.[11]

1.1.4   Compositionality

The utility of a composite often depends on the utilities of its components. For example, the utility of a basket of goods may depend on the utilities of the goods. This section introduces the compositionality of a property's application to composites and explains its relation to the separability of the property's application to the composites. It treats utility as a quantitative property that a proposition's realization may have to various degrees.

A property's application to a composite is *compositional*; that is, it decomposes into its applications to the composite's parts if and only if the property's application to the composite is a function of the property's application to its parts, taken in their order if the composite orders them. Because compositionality may hold for one division but not for another division of a whole into parts, a complete claim of compositionality specifies not only a property but also a division of a whole into parts. A complete claim explicitly states the division into parts that a property's compositionality assumes if context does not settle the division. Compositionality for a property and a division of composites, for some range of composites and a context, asserts the existence of a function that obtains the property's application to a composite from the property's application to the composite's parts. For utility's application to a proposition, a principle of compositionality may specify a method of dividing the proposition into parts. One principle of compositionality, for a type of utility, may divide a conjunction into its conjuncts.

---

[11] The mutual separability of two variables does not entail that the utility of the first's value and the utility of the second's value contribute independently to their combination's utility. Mutual separability holds trivially for a single vector (x_1, y_1). It fails only for an order of multiple vectors and subvectors. So the argument variables of $U(x, y)$ are mutually separable given the unique vector of variable values (x_1, y_1). However, independent contribution may fail for a single vector. The utilities of the values of two variables may not contribute independently to the vector's utility. The values of the variables may be complementary, as in the case of a right glove and a left glove, despite the variables' mutual separability given their unique vector of values.

For a kind of utility U applied to an ordered composite $(x_1, x_2, \ldots, x_n)$, compositionality claims that $U(x_1, x_2, \ldots, x_n) = F(U(x_1), U(x_2), \ldots, U(x_n))$. This equation does not assert that the utilities of a whole's parts settle the whole's utility. It recognizes that the order of the parts may contribute to the whole's utility. The composite may present events in their temporal order. The events' temporal order may affect the composite's utility. Taking an introductory logic course before an advanced logic course is better than taking the courses in the reverse order. The order of courses in a sequence settles the order of arguments in the function that obtains the sequence's utility from the courses' utilities. If composites order parts, compositionality uses a function that incorporates their order. Utility's compositionality for pairs (x, y) entails that $U(x)$ and $U(y)$ in this order settle $U(x, y)$ so that no variation in conditions that does not affect $U(x)$ and $U(y)$ affects $U(x, y)$.

According to the principle of *interchange of equivalents*, for a property applying to objects and the composites they form, replacing a composite's part with an object of the same value yields a new composite with the same value as the original, provided that the new part does not interact with remaining parts to generate an additional part. Take a composite with several parts, and assume that a utility function applies to each part as well as to the whole composite. What happens to the composite's utility after replacing some part with an object of the same utility? If the principle of interchange of equivalents holds, then the composite's utility stays the same, assuming that the exchange is not productive. For example, suppose that a shopper's basket of goods includes yogurt, and the shopper exchanges the yogurt for another brand of the same utility. If the principle applies, the shopping basket's utility before the exchange is the same as after the exchange.

Compositionality and interchangeability both treat relations between a property's application to wholes and its application to parts. In fact, interchangeability is equivalent to compositionality; that is, it is necessary and sufficient for compositionality, given a common range of composites.[12]

Compositionality and separability live in the same neighborhood. Compositionality governs a property's application to composites and their components. Separability governs orders of composites and their components, and a property such as utility may generate the orders. Both compositionality and separability apply to variables, for their ranges of values. Both are trivial if each variable has just one value. Types of complementarity oppose both compositionality and separability.

---

[12] Westerståhl and Pagin (2011) define compositionality and note its equivalence with interchange of equivalents, given observance of domain restrictions.

A difference in quantification distinguishes compositionality and separability. A utility function $U(x, y)$'s compositionality and its mutual separability for x and y entail, respectively, generalizations with quantifiers of different scope. According to compositionality, for all $x, y, U(x, y) = F(U(x), U(y))$ for some F. According to mutual separability, for all y, for all x and x', $U(x) \geq U(x')$ if and only if $U(x, y) \geq U(x', y)$, and similarly for all x, for all y and y'. Compositionality and separability differ over arrangements of factors to fix and to let vary.

In the utility function $U(x, y)$, the first variable's separability from the second does not suffice for compositionality. If x is separable from y, then $U(x, y) = F'(U(x), y)$ for some function F'. If two values of y have the same utility, the utility of a value of x and a value of y may depend on y's value and not on the utility of y's value despite the separability of x from y. Then, it is not the case that $U(x, y) = F(U(x), U(y))$ for some function F. Compositionality follows from the variables' separability from each other, however. If x and y are separable from each other, then $U(x, y) = F(U(x), U(y))$ for some function F. More generally, in a utility function for vectors, separability of each argument variable from the others implies compositionality. If the order of each argument's values, holding the other arguments fixed, agrees with the vectors' order, then the utility of a vector is a function of the utilities of the arguments' values.[13]

In the example, the implication is trivial if no values of x have equal utility, and no values of y have equal utility. To make the implication nontrivial, suppose that y_1 and y_2 have the same rank and so the same utility, let (x_1, y_1) and (x_1, y_2) have the same rank, and let (x_2, y_1) and (x_2, y_2) have the same rank. Furthermore, let x_1 come below x_2, and let (x_1, y_2) come below (x_2, y_1). Then, x and y are mutually separable. Substituting equivalents, y_2 for y_1, to move from (x_1, y_1) to (x_1, y_2) does not change the composite's utility. The principle of interchange, which is equivalent to compositionality, holds.

Although each variable's separability from the others implies compositionality, compositionality does not imply each variable's separability from the others (even if the order of vectors defines the order of subvectors). A

---

[13] Suppose that in $U(x, y)$, the variable x is separable from y, and the variable y is separable from x. That is, the order of (x, y)'s values agrees with the order of x's values however y's value is fixed, and the order of (x, y)'s values agrees with the order of y's values however x's value is fixed. Then, $U(x, y) = F'(U(x), y)$ for some function F' given x's separability from y, and $U(x, y) = F(U(x), U(y))$ for some function F given y's separability from x. The function F increases with increases either in $U(x)$ or in $U(y)$. In a utility function with more than two argument variables, despite separability of one argument variable from the others, aggregation of the other argument variables, and so the whole set of argument variables, may not be compositional. Separability of each argument variable from the others ensures compositionality, however. Broome (1991: 69) states this point.

property's compositionality does not entail the separability of components of composites with the property because separability concerns orders. The property may not impose an order on the composites and their components. Meaning is compositional, supposing that Frege is right about meaning; the meaning of a sentence is a function of the meaning of its parts. However, sentence meanings do not order sentences, and word meanings do not order words. No agreement between the order of subjects and the order of sentences obtains when a sentence's predicate is held constant but its subject is allowed to vary. The words of a sentence are not separable with respect to any function representing order.

Even if a compositional property imposes an order, compositionality does not entail separability. Consider an order of pairs (x_1, y_1), (x_2, y_2), (x_1, y_2), (x_2, y_1) from lowest to highest. The variable x is not separable from y (given any order of elements). With y_1 fixed, the agreeing order for the first element is $x_1 < x_2$, but with y_2 fixed, the agreeing order for the first element is $x_2 < x_1$. No single order of x_1 and x_2 holds however y's value is fixed. Although in $U(x, y)$ the variable x is not separable from the variable y because it is not separable from y in the order of vectors (x, y) that the function represents, $U(x, y)$ may be a function of $U(x)$ and $U(y)$, and so compositional. In the example, let the order of pairs give the pairs their utilities, let the order of elements by subscript give the elements their utilities, and replace the elements with their utilities to obtain a function from their utilities to the pairs' utilities. Imagine that $U(x_1) = 1$, $U(x_2) = 2$, $U(y_1) = 1$, $U(y_2) = 2$, $U(x_1, y_1) = 1$, $U(x_2, y_2) = 2$, $U(x_1, y_2) = 3$, and $U(x_2, y_1) = 4$. Then, construct F so that $F(1, 1) = 1$, $F(2, 2) = 2$, $F(1, 2) = 3$, and $F(2, 1) = 4$. As a result, $U(x, y) = F(U(x), U(y))$, which establishes compositionality.

Other cases in which compositionality and separability come apart begin with compositionality and show how separability fails. If a pair's utility equals a ratio of its elements' utilities, then the pair's utility is compositional, but the order of pairs inverts the order of the denominator's values given a fixed numerator, and so the denominator is not separable from the numerator. Next, suppose that x and y are variables with varying possible values. $U(x, y) = F(U(x), U(y))$, so compositionality holds. Suppose that $U(x_1, y_1) = F(U(x_1), U(y_1)) = 1$, $U(x_1, y_2) = F(U(x_1), U(y_2)) = 2$, $U(x_2, y_1) = F(U(x_2), U(y_1)) = 2$, and $U(x_2, y_2) = F(U(x_2), U(y_2)) = 1$. The order of y_1 and y_2 agreeing with the order of pairs changes as U's first argument shifts from x_1 to x_2. So, the second argument place is not separable from the first.

To illustrate concretely compositionality without separability, imagine that the utility of wealth and security together is a function of the utility of wealth and the utility of security. That is, no utility-divergent pairs of a level of wealth and a level of security have levels of wealth with the same utility and levels of security with the same utility. Suppose that if security is set low, then increasing the utility of wealth from a high level to a slightly higher level in a way that triggers publicity decreases overall utility by increasing the probability of kidnapping. In contrast, if security is set high, then increasing the utility of wealth increases overall utility. The utility of a composite may be a function of its components' utilities without being a separable function of their utilities, that is, a function in which each argument is separable from the others.

Separability entails a type of independence. Does a weaker type of independence combine with compositionality to yield separability? Independence may hold between variables, variables' values, or their utilities besides holding between orders of variables' values, as with separability. Some types of independence involve constancy as conditions change, and some involve the absence of interaction. Consider a set of variables, a subset of them, and vectors and subvectors that list values of the variables in the set and in the subset, respectively. Suppose that preferences order the vectors and subvectors and that for a single subvector, its utility is the same given any values of its complement. This is a type of independence involving constancy for utilities of variables' values. Suppose that it holds for all subvectors of values of variables in the subset. Then, the order of subvectors of the subset's values is the same for all values of the subset's complement. This is conditional independence of the order. Does compositionality and conditional independence imply separability? No, separability of the subset of variables from its complement requires that the order of subvectors agree with the order of vectors for all values of the subset's complement. Even though the order of subvectors is independent of its complement, it need not agree with the order of vectors.

To illustrate, suppose that one variable is for a world's past and another is for a world's future (including the present). Compositionality for utility entails that when the past is fixed, a world's utility is a function of its future's utility. It prohibits assigning different utilities to two composites whose elements have the same utility profiles. With the past fixed, futures of the same utility yield worlds of the same utility. A future's utility settles a world's utility. Nonetheless, the direction of a future's influence on a world's utility may vary from future to future. For one future, its utility may increase the world's utility, whereas for another future, its utility may decrease the world's utility. This happens if the utility of the past is two, the utility of the first future is one, the utility of the second future is two, and the future's utility adds to the utility of the past if the total of past and future utilities is three or less and subtracts from the utility of the past if the total of past and future utilities is more than three because of penalties for exceeding a limit on their sum. Conditional independence makes the direction of influence uniform for all ways of fixing the past. However, separability requires more. It requires that with the past fixed, no matter how, the order of futures agrees with the order of worlds. That a world's utility is a function of its parts' utilities follows from the separability of each part from the others but does not entail that in a utility function for the world, each argument is separable from the others.

Compositionality simplifies evaluation of options, as does separability. It makes a composite's utility calculable from the utilities of its parts. However, compositionality is not enough to simplify deliberations given incomplete information. Suppose that utility attaches to worlds, their pasts, and their futures, and the utility of a world is a function of the utility of its past and its future, so that holding the past fixed, the utility of a world's future settles the world's utility. That is, $U(p, f) = F(U(p), U(f)) = F'(p, U(f))$ for some F and F'. Letting p have the actual past as its value, the utilities of futures settle the utilities of worlds. Given conditional independence, a future's utility influences a world's utility in a uniform direction. However, the direction of influence may depend on the past. Separability rules out that dependence and so allows for ignorance of the past. Given that the future is separable from the past, $F(U(p), U(f))$ is strictly monotonically increasing with respect to $U(f)$. Strict monotonicity puts the order of futures in step with the order of worlds for any value of p. Given separability, an evaluator need not know the past to know that the order of futures agrees with the order of worlds. This simplifies options' evaluations.

1.2   Establishing separability

Various principles of separability govern the various types of utility. Chapter 3 argues for a fundamental form of separability, and Chapters 2 and 4–6 argue for derived forms of separability. This section assembles general strategies for arguments supporting separability.

1.2.1   Norms

Separability is a relation among orders of composites or utility functions representing the orders. Rationality requires some forms of separability, as later chapters argue. The separation of the future from the past is a normative requirement. For a rational ideal agent and type of utility, a world's utility separates into the past's utility and the future's utility. The separation encourages deliberations to be forward looking and to ignore sunk costs.

Utility is rational degree of desire in an ideal agent. For a type of desire and utility, showing that rationality requires an ideal agent's degrees of desire to be separable also shows that utilities, a rational ideal agent's degrees of desire, are separable. Utilities obey all principles of rationality including principles of separation. Utilities do not obey the principles by definition alone but also because of rationality's regulation of degrees of desire. A type of utility's separability is normative given that an ideal agent's rationality is responsible for it.

Methods of establishing a form of separability depend on whether the separability is normative or empirical. If it is empirical, then gathering evidence is a basic method of supporting it. For example, evidence supports the combined gas law, according to which the pressure of a gas is a separable function of the gas's temperature and the reciprocal of its volume. The scientific method directs justification of an empirical principle of separability. If separability is normative, then its support comes from normative principles and resolutions of cases. For example, the moral theory, utilitarianism, advances a normative principle claiming that utility for a group of people is a separable function of utilities for the group's members. Moral principles and resolutions of cases provide its support. This book treats normative separability, in particular, separability that rationality requires. Taking separability as a norm demands philosophical rather than empirical justification.

Suppose that rationality requires a type of separability for a utility assignment. Then, the norm is an a priori truth and every set of assumptions implies it. Giving assumptions that imply it does not establish the norm. An effective argument for it states assumptions that explain the norm, perhaps normative assumptions that logically or mathematically imply the norm.

Do representation theorems offer a method of establishing separability? A representation theorem for a preference order over composites demonstrates the possibility of representing the order with a utility function that is unique given a zero point and a unit. The appendix to this chapter reviews famous representation theorems that Gérald Debreu and William Gorman prove, stating that some types of ordinal separability of variables suffice for an additive utility-representation of the order of the variables' values. A sufficient type is called *additive separability* and obtains if each set of variables is separable from its complement.

Representation theorems use empirical and normative assumptions that regiment the preferences that utilities represent. A unique representation of an order of composites, given a utility scale, exists only if the order has suitable complexity and structure. An additive representation's existence and uniqueness require multiple composites and require putting on the same scale utility functions that represent the order of the composites and the order of the values of variables generating the composites. Some structural assumptions are empirical rather than normative. For two variables x and y specifying amounts of commodities, commutativity, a necessary condition of an additive representation, is a norm for utility: $U(x, y) = U(y, x)$. A rational ideal agent complies with the norm and so facilitates construction of an additive utility function that represents her preferences. However, the composites' generation from variables with multiple values, a requirement for the representation, is not a norm. Whether it holds is an empirical matter. The representation theorems that the appendix reviews take separability as an empirical fact about preferences that grounds a utility-representation of preferences with certain structural properties such as additivity. They do not offer a means of justifying separability as a norm.

Normative principles of separability typically use simplifying assumptions that create a normative model. Both philosophy and economics use models. Typically, economic models of choice serve empirical goals, whereas philosophical models of choice serve normative goals. Even when models in economics and in philosophy treat the same normative topic, they differ because the disciplines have different objectives. Philosophy does not exclude a factor from a model because it makes the model difficult to apply to practical problems, whereas economics does. Because the two disciplines' models serve different purposes, they are subject to different standards. This book justifies norms of separability within a philosophical model that includes the idealization that agents are rational and cognitively unlimited, and also the restrictive assumption that agents have basic intrinsic attitudes attaching to proper parts of worlds. The idealizations and restrictions limit the book's normative principles of separation.[14]

Utilitarianism creates a philosophical normative model. The model assumes interpersonal comparisons of utility for alternative courses of action affecting the people in a population. Being philosophical, the model accepts interpersonal comparisons of utility despite the practical difficulty of making such comparisons. In light of judgments about right acts, it advances the

---

[14] It is convenient to take a model as a possible world or set of possible worlds meeting a group of assumptions. Investigations of the model reveal the assumptions' implications. A typical normative model for decision theory is a possible world that meets a group of assumptions about agents, their decision problems, and their circumstances. Applications of principles of rationality within the model reveal the assumptions' normative implications.

principle that an act is right if and only if it maximizes collective utility. The normative principle assumes the additive separability of an act's collective utility; it is the sum of the act's utilities for the population's members. This separability is also normative unless part of collective utility's definition.

Norms of separability for personal preferences, desires, and utilities come in varying strengths. A normative principle for preferences requires the possibility of a representation that obtains the utility of a world from a separable function of the utility of the world's past and the utility of the world's future. A stronger normative principle requires, for a type of desire and utility taken as rational degree of desire, separation of a world's utility into its past's utility and its future's utility. Chapter 4 argues for the stronger normative principle. In general, the chapters presenting norms of separability target utilities taken as rational degrees of desire and not utilities defined using preferences.

What type of separability does simplification of choices require? At a minimum, it demands, for options' worlds and their preference ranking, the separability of a world part evaluated from a world part put aside, for example, assuming evaluation of options by their futures, the separability of the future from the past. However, the additivity of a world's utility, for a type of utility and division of the world, and the more demanding type of separability such additivity entails, is not superfluous.

Consider an analogy. Although utilitarianism's goal is a collective ranking of collective options, the introduction of collective utilities in addition to the collective ranking is not superfluous. The collective utilities generate the collective ranking. The options' rankings by individuals do not suffice for the collective ranking. The rankings by individuals yield, following the *Pareto principle*, according to which unanimous preference settles collective preference, only an incomplete collective ranking. Interpersonal utilities for individuals, by generating collective utilities, complete the collective ranking.

Similarly, although deliberations rank options, introducing utilities of option's worlds is not superfluous. The worlds' utilities generate a ranking of the worlds that in turn generates a ranking of the options. Ranking worlds' futures and ranking their pasts, the analog of two individuals ranking collective acts, is not enough to generate the ranking of worlds. The rankings of pasts and futures yield, following an analog of the Pareto principle, only an incomplete ranking of worlds. Utilities of pasts and futures, by generating utilities of worlds, complete the ranking of worlds. For example, take two worlds. In the first, the past is better than in the second, and in the second, the future is better than in the first. Neither world is Pareto superior to the other. However, given utilities for each past and for each future, and the additivity of worlds' utilities, a ranking of the worlds emerges. Utility's additivity makes a contribution beyond preference's separability.

Various types of separability apply to a set of parts. Additive separability of a world's components, with respect to preference, grounds addition of components' evaluations to obtain a world's utility. If a world's utility is additive (for a type of utility), then subtracting a part's evaluation is equivalent to a scale change for the world's utility. Taking world parts as variables, the order of a variable's values agrees with the order of their utilities' sum after fixing the values of the other variables. Although separation of a world's part from its complement is sufficient for separable evaluations that simplify choices, utility's additivity for some world parts is not superfluous. Its additivity for realizations of basic intrinsic attitudes grounds the definitions of temporal, spatiotemporal, and causal utility, and those types of utility yield an evaluation of a world's part that is separable from an evaluation of its complement. Utility's additivity for world parts grounds the separability of preferences that simplifies choices.

1.2.2   Fundamental separability

Some quantities are additive because of conventions of measurement. Weight, operationally defined, adopts as a *concatenation operation*, to represent with addition, combining two objects to obtain another object. Adopting a concatenation operation for a quantity's measurement establishes the additivity of the quantity measured with that operation. Weight is additive because the quantity's measure adopts combination as a concatenation operation. The weight of a combination of objects, according to the measure, is the sum of the weights of the objects it combines. The measure, the weight function, makes weight additive.

An alternative view of weight takes it as a theoretical quantity manifest in various phenomena such as pull on springs and tipping of balances. Assuming additivity in one manifestation may count against additivity in another manifestation. Weight according to balances, assumed to be additive, may not be additive according to springs. Failures of additivity in some manifestation may suggest that weight changes in combinations or that combinations produce new objects with weight. Representation theorems show how to use additivity to measure a quantity in some manifestation under assumptions, such as constancy in combinations, but do not define the theoretical quantity. Operationally defined using balances, an object's weight given a unit is relative to comparisons of a set of objects, including combinations, on a balance. However, taken as a theoretical quantity, its weight given a unit is not relative this way.[15]

Two conceptions of utility differ as the two conceptions of weight differ. According to one, utility's additivity (for a type of utility, range of composites, and their division) is a consequence of a conventional concatenation operation for composite objects. According to the other, utility is a theoretical quantity measurable in various ways. A concatenation operation, assuming its accuracy, offers a method of measuring but not defining the theoretical quantity. The first conception defines utility as a construct from preferences meeting certain constraints and makes a composite's utility given a unit relative to preferences over a set of composites. The second conception, taking utilities to be strengths of desire, makes utility given a unit conceptually independent of preferences and constraints on them, and nonrelative to preferences over a set of composites.

How may an argument establish additivity for a theoretical quantity such as degree of desire without establishing it as a convention of measurement, as does a typical operational definition of the quantity? A strategy uses representation theorems the appendix describes to show that if an order of composites meets certain conditions, including some separability conditions, then the order has an additive representation. However, this strategy, which is not general because of the conditions it imposes, uses the separability of an order of composites to establish the possibility of an additive representation of the order and does not establish the additivity of the theoretical quantity that generates the order.

Probability theory illustrates another method of deriving one type of separability from another. Support for additivity of probability, taken as rational degree of belief in ideal agents, may equate probabilities with additive quantities, such as ideal betting quotients. Degrees of belief yield betting quotients, and some groups of bets made with nonadditive betting quotients are vulnerable to sure losses. Ideal betting quotients are additive; the quotients for bets on disjunctions of exclusive propositions decompose into the quotients for bets on the disjuncts. Also, degrees of belief should match strengths of evidence, and the strength of evidence for a disjunction of exclusive propositions is a sum of the strengths of evidence for the disjuncts. So in an ideal agent, degrees of belief, if rational, inherit their additivity from strengths of evidence. Such arguments for probability's additivity take some type of

---

[15] Representation theorems explain measurement of probability and utility in ideal cases even if, for reasons that Meacham and Weisberg (2011) review, they do not offer good definitions of probability and utility.

additivity as given and derive probability's additivity from it. They show that probabilities are additive by showing that probabilities match other additive quantities, such as ideal betting quotients or strengths of evidence.

These arguments do not take probability's additivity as fundamental, so they push back the problem of establishing additivity. Suppose that probabilities are fundamentally additive. For a rational ideal agent, the degree of belief that a disjunction of incompatible propositions holds equals the sum of the degrees of belief that the propositions hold. Then, additivity receives support as an intuitive generalization of intuitions about examples. Support may also show that probability meets necessary conditions of additivity, such as separability. Given that the probability of a disjunction of exclusive propositions equals the sum of the probabilities of the disjuncts, the probability of the first disjunct is separable from the probability of the second disjunct. No matter how the probability of the second disjunct is fixed, increasing the probability of the first disjunct increases the probability of the disjunction. An order of disjunctions of mutually exclusive propositions is separable because the order of disjunctions with a common disjunct agrees with the order of the other disjuncts. Showing this separability, a necessary condition of probability's additivity, involves examining cases and generalizing from them. Showing that addition has no structural properties that the relations among degrees of belief lack, if the agent is ideal and the degrees of belief are rational, motivates measurement's taking disjunction of mutually exclusive propositions as a concatenation operation for probability. Probabilities are additive given this justified, as opposed to conventional, concatenation operation.

Arguments for a type of utility's additivity depend on whether it, and the separability it presumes, is fundamental or derived. An option's utility equals its expected utility, and an option's expected utility is a sum of the probability-utility products for the option's possible outcomes. Therefore, an option's utility is a sum of the utilities of the chances for its possible outcomes. The next chapter argues for this additivity of comprehensive utilities. It derives from probability's additivity.

Chapter 3 argues for intrinsic utility's separability and additivity. Its separability and additivity are fundamental, and so their support uses the independence of intrinsic utility's order of realizations of basic intrinsic attitudes. As Chapter 3 shows, if an agent has basic intrinsic desires for health and for wisdom, and realizing the first desire has greater weight, then it has greater weight no matter which other basic intrinsic attitudes are realized. This independence grounds intrinsic utility's separability and additivity. Independence of reasons for basic intrinsic attitudes supports the separability of the intrinsic utilities of combinations of their realizations. The argument uses a type of independence to support a type of independence but is not question begging because of the difference between independence of reasons and independence of utilities.

This chapter defined separability for a preference order and a utility function in a way that makes separability apt for simplifying choices. It also showed that complementarity opposes separability and that separability implies compositionality. Because later chapters argue for the separability of various types of utility, it identified suitable methods of arguing for separability.

1.3   Appendix: Theorems

Famous theorems treat an order's separability. They characterize relations between types of separability and show that extensive separability of variables in a set supports an additive representation of an order of vectors of the variables' values.

Suppose that a set of variables generates a rich array of vectors with an order (which may define orders of subvectors). In the set of variables, some subsets may be separable from other sets. In the set, *weak separability* holds if and only if each variable is separable from the others. As Section 1.1.4 notes, weak separability entails compositionality. *Strong separability* holds if and only if every set of variables is separable from its complement. Strong separability is necessary and sufficient for an additive representation (unique up to positive affine transformations) of the order of the vectors that the variables generate. When and only when an additive representation is possible, the variables are *additively separable*: using utilities to represent the order of occurrences at a location, for a vector the sum of occurrences' utilities represents the vector's place in the vectors' order. *Crosscutting or overlapping separability* (which has a complex definition) suffices for an additive representation because it implies strong separability. Table 1.2 uses arrows to indicate entailments. Broome (1991: 70, 82–89) states the theorems of separability and sketches proofs.[16]

As Chapter 2 introduces utility, a utility function indicates strength of desire. The function represents preferences, but preferences do not define

---

[16] Gorman ([1968] 1995: chap. 12) shows that the overlapping or crosscutting condition establishes additive separability. Krantz et al. (1971: sec. 6.11) state Debreu's theorem: strong separability implies additive separability. Keeney and Raiffa ([1976] 1993: chap. 3) state Debreu's and Gorman's theorems on additive separability.

Table 1.2 *Properties of an order of composites*

Additive		Weak	
Separability $\leftrightarrow$	Strong Separability $\rightarrow$	Separability $\rightarrow$	Compositionality
	$\uparrow$		
	Overlapping Separability		

the function. Given utility's definitional independence from preferences, utility's additivity means more than the possibility of an additive utility-representation of preferences concerning values of a set of variables. The utility function's additivity means that the utility of a vector of values equals the sum of the utilities of the values (given a common scale for their utilities). The utilities of vectors and subvectors may generate preference orders for the vectors and subvectors. A unique additive utility-representation of a rational ideal agent's preferences, given a choice of utility scale, reveals utilities in the sense of strengths of desire (on the same scale) if the utilities are additive.

A strategy for establishing an additive utility-representation of preferences over a set of composites uses, first, the theorem of overlapping sets to extend separation of some subsets of variables to separation of all subsets from their complements and, second, the theorem of strong separability to transform separability of all subsets from their complements into additive separability. However, the strategy yields only an additive utility-representation and not the additivity of utility taken as strength of desire.

Utility's additivity entails that for some wholes and divisions into parts, the degree of desire for a whole equals the sum of the degrees of desire for the whole's parts. Utility's additivity is not necessary for an additive utility-representation of preferences. A preference order may be additively separable without utilities being additive, if the order has just one vector, because in this case separability holds trivially whereas additivity is substantive. Also, if the order of vectors defines the order of subvectors, the order of a variable's values according to utility may run counter to the order of vectors so that in a utility function for the vectors, the variable is not separable from its complement, and utility is not additive. Establishing additivity for utilities taken as quantitative representations of desires differs from using the additive separability of a preference order to ground an additive utility-representation of preferences. In rational ideal agents, additive separability is easier to establish for a preference order than is additivity for a utility function because utility's additivity entails an additively separable preference order, whereas an additively separable preference order does not entail utility's additivity. An additively separable preference order of composites has an additive utility-representation. However, the representation may not accurately represent utility taken as strength of desire and so may not establish utility's additivity.

In some quantitative functions, arguments are additively separable although not additively aggregated. According to probability theory, the probability of a pair of independent events equals the product of the events' probabilities: $P(A \ \& \ B) = P(A)P(B)$ if A and B are independent events. Take this equality as a principle of probability governing informationally independent events rather than as a definition of independent events. Then, for independent event-variables A and B, composites formed with binary conjunction, and a probability function P that represents comparative probability, the multiplication principle makes $P(A \ \& \ B)$ an additively separable function, but not an additive function, of $P(A)$ and $P(B)$.[17]

Using the theorems of separability to argue for utility's additivity faces two shortcomings. First, although utility's additivity holds generally, the assumptions under which the theorems show that a preference order has a unique additive utility-representation do not hold generally. Second, demonstrating that given the assumptions, a preference order has a unique additive utility-representation does not establish utility's additivity for the composites in the order. Therefore, Chapters 2–6 do not use the theorems of separability to argue for utility's additivity.

---

[17] A common *return-risk* method of evaluating an investment uses the investment type's coefficient of variation s/m, where s is the standard deviation for returns from the investment type, and m is the mean of returns from the investment type, or the investment's expected return. The smaller the coefficient of variation, the better the investment. Hence, the greater m/s, the better the investment. The reciprocal of the coefficient of variation yields a measure of an investment's value that uses assessments of an investment with respect to return and with respect to risk (taken in a technical sense). An investment's value, according to the evaluation, is proportional to return given a fixed risk. Also, an investment's value is inversely proportional to risk given a fixed return. Multiplying return and the reciprocal of risk yields a quantity proportional to an investment's value. The reciprocal of risk is separable from return, and return from the reciprocal of risk. Multiplication is a function besides addition with separable factors (although it may be transformed into addition using logarithms). Gorman ([1968] 1995: chap. 14) notes that addition and its transformations are pretty much the only associative functions that make factors separable. Addition is the most common way of representing factors' separability.

CHAPTER 2

Expected utility

A wise gambler evaluating a bet divides the bet's possible outcomes into winning and losing. After evaluating these possible outcomes, the gambler weights their evaluations according to their probabilities and combines their weighted evaluations to obtain the bet's evaluation. Expected-utility analysis follows this commonplace pattern. To calculate an option's comprehensive utility, it identifies mutually exclusive and jointly exhaustive possible outcomes of the option. Then, it separates an option's utility into utilities that attach to chances for the possible outcomes. These utilities are probability-weighted utilities of the possible outcomes; the utility of a chance for a possible outcome equals the product of the possible outcome's probability and utility. An option's utility equals the sum of the chances' utilities. So, the analysis's principle of separation, the expected-utility principle, states that an option's utility equals its expected utility, a probability-weighted sum of the utilities of the option's possible outcomes.[1]

The expected-utility principle asserts that for a certain division of a whole into parts, the utility of the whole is a sum of the utilities of the parts. The principle rejects holism for an option's utility. Arguments for the principle support addition of utilities of chances an option offers. This chapter explicates evaluation of an option by expected utility. Expected-utility analysis's additive separation of an option's comprehensive utility into parts is a model for other separations of utilities into parts. Because this chapter's formulation is general and applies to comprehensive, intrinsic, temporal, spatiotemporal, and causal utility, it also grounds other separations of utilities when information about an option's outcome is incomplete. Other chapters argue for additional forms of utility analysis using this chapter's precise formulation of expected-utility analysis. Refining this familiar form of utility analysis strengthens their arguments.

---

[1] This principle concerning an option's utility, which Weirich (2010b, 2010c) elaborates, follows Jeffrey's ([1965] 1990: 78) characterization of a proposition's desirability.

The normative model for the expected-utility principle assumes rational ideal agents. For these agents, an option's utility equals its expected utility. The principle rests on norms of rationality as well as the definitions of probability and utility. Also, this chapter's expected-utility principle, for comprehensiveness, defines an option's outcome using a possible world. The principle's analysis of an option's utility is called world-Bayesianism and belongs to Bayesian decision theory because it uses subjective or information-sensitive probabilities to define an option's expected utility. The chapter explains this type of probability, utility, and the standard of expected-utility maximization for a decision's evaluation. Expected utility's place in the standard provides a test of the expected-utility principle. For an option's utility to equal its expected utility, maximization of expected utility must be equivalent to maximization of utility.

Instead of defining probabilities and utilities as constructs from preferences, this chapter defines probability as rational strength of belief and utility as rational strength of desire. It explains and motivates this realist treatment of probability and utility and argues for the expected-utility principle it yields. The chapter's presentation of the realist view is brief because the view is common (especially in psychology, management, and finance) and because Weirich (2001) presents it at length. Also, adopting an alternative, constructivist view of probability and utility does not upset the project of simplifying choices. It just makes the expected-utility principle hold by definition rather than as a norm. Using definitions instead of norms to ground separability still allows separability to simplify choices.

The sections to follow cover probability, utility, outcomes, the expected-utility principle, its generality, support for the principle, and rational choice. The expected-utility principle advances a basic type of separability according to chances, and Chapter 3 advances another basic type of separability according to goals. Chapters 4–6 use the two basic types of separability to derive other types that simplify choices.

2.1 Probability

Physical probabilities depend on physical facts, such as relative frequencies. They are insensitive to changes in information (unless they are about information). For example, the physical probability of getting a black ball when randomly drawing a ball from an urn depends on the percentage of black balls in the urn. It does not depend on information about that percentage.

Evidential probabilities are sensitive to information, that is, evidence. The probability that Shakespeare wrote *Hamlet* varies with the evidence for the play's authorship. It is high unless new evidence arrives showing that someone else, perhaps Francis Bacon, wrote the play. In contrast, the physical probability that Shakespeare wrote the play is either 0 percent or 100 percent and is constant as evidence changes.[2]

Degrees of belief form the foundation of evidential probabilities. They are propositional attitudes and attach to propositions representing states and outcomes. I assume that an ideal agent in an ideal decision problem has precise degrees of belief concerning relevant propositions but do not assume that an agent's evidence in all cases demands a sharp degree of belief concerning every proposition or that a rational ideal agent assigns a precise degree of belief to every proposition.[3] The strength of a belief depends on its comparison with a unit, by convention a belief of maximum strength such as a belief in a simple logical truth. Agents who reflect on their degrees of belief can often identify them at least roughly. People who consider the question generally know that they have a degree of belief of 50 percent that a toss of a fair coin will yield heads. A person's betting rates also reveal her degrees of belief. She typically has a degree of belief of 50 percent that a toss of a fair coin will yield heads if she is willing to buy or sell for $0.50 a gamble that pays $1 if the coin toss yields heads and $0 otherwise. Degrees of belief are implicitly defined by their role in an agent's mental life and by their causes and effects. Because they are defined independently of the probability axioms, their conformity to the axioms is a norm rather than a definitional truth.[4]

Evidential probability is rational degree of belief. More precisely, it is rational degree of belief for an ideal agent who knows all logical and mathematical truths, has no cognitive limits, is in ideal circumstances for formation of degrees of belief, and does not have any practical goals in conflict with her epistemic goal of having any degree of belief that she

[2] In some works, evidential probability means information-sensitive probability in the spirit of Kyburg (1974). Kyburg grounds information-sensitive probabilities in statistical data only. In contrast, evidential probability in this book's sense allows for probabilities grounded in nonstatistical evidence.

[3] Objective Bayesians hold that evidence settles precise degrees of belief. Elga (2010) argues that unsharp probabilities may yield an incoherent series of choices so that rationality requires sharpness for practicality's sake. If he is right, then agents have pragmatic reasons for sharpness even if they lack evidential reasons for sharpness. This chapter maintains neutrality on rationality's requirements concerning precision in degrees of belief.

[4] Christensen (2004: chap. 5) argues against explicitly or implicitly defining degrees of belief using preferences alone. The mental role of degrees of belief connects them with mental states besides preferences. For example, as Christensen (2004: 129, 131) notes, the greater one's degree of belief that a burglar is in the house, the greater is one's fear.

attaches to a proposition conform to the strength of her evidence for the proposition's truth. The idealization ensures that rational degrees of belief comply with the laws of probability. In a rational ideal agent, degrees of belief are nonnegative, tautologies enjoy maximum degree of belief, and the degree of belief that a disjunction of incompatible propositions holds equals the sum of the degrees of belief that the disjuncts hold.[5] One exception to the laws may arise. Even given the idealization, rationality may not demand that the propositions to which degrees of belief attach form a Boolean algebra. Perhaps a rational ideal agent may fail to have a degree of belief concerning the conjunction of two propositions for which she has degrees of belief. Consequently, the idealization ensures that evidential probability obeys the standard laws of probability, except laws of probability's existence over propositions forming an algebra; a rational ideal agent's degrees of belief obey the regulative probability axioms. Various arguments aim to establish this conformity. Although theorists debate its explanation, the conformity itself is not subject to serious doubt, and this book just takes it for granted.

The probabilistic structure of a rational ideal agent's degrees of belief simplifies their measurement. For example, if concerning a coin toss, a rational ideal agent's degree of belief that heads will come up equals his degree of belief that tails will come up, then his degree of belief that heads or tails will come up is twice his degree of belief that heads will come up. A norm for a rational ideal agent makes disjunction of incompatible propositions a concatenation operation that yields instruments for measuring degrees of belief.

As in Krantz et al. (1971: chap. 5), a representation theorem may show that comparative belief meeting certain structural and normative conditions has a unique representation by a probability function, which by definition meets the axioms of probability. A theorem, if constructive, shows how to derive degrees of belief from comparative belief and so describes a measurement procedure. However, the theorem does not establish that degrees of belief should satisfy the probability axioms because their representation by a probability function just assumes that they do. The norms that govern comparative belief ensure the possibility of a representation by a probability function but do not establish probabilistic norms for degrees of belief.

Suppose that some system of comparative belief ranks $(s \lor \sim s)$ above s, s with $\sim s$, and $\sim s$ above $(s \ \& \ \sim s)$. These comparisons meet all norms for

---

[5] Because I treat only finite cases, I do not take a stand on the axiom of countable additivity for rational degrees of belief.

comparative belief. A simple representation theorem states that for all propositions p and q in the order's field, for a unique probability function P, p ranks above or with q if and only if $P(p) \geq P(q)$. According to the function, $P(s \vee \sim s) = 1$, $P(s) = P(\sim s) = 0.5$, and $P(s \mathbin{\&} \sim s) = 0$. This result does not establish that for propositions in the order, degrees of belief should satisfy the axioms of probability. Although the comparisons in the example have a representation by just one probability function, they also have representations by nonprobabilistic functions, such as Q, according to which $Q(s \vee \sim s) = 1$, $Q(s) = Q(\sim s) = 0.4$, and $Q(s \mathbin{\&} \sim s) = 0$. A set of comparisons' representation by a unique probability function does not rule out its representation by nonprobabilistic functions. Assuming that degrees of belief satisfy the axioms establishes the probabilistic representation's fidelity to degrees of belief but does not establish the axioms.

Standard decision principles assume that an agent has access to evidential probabilities used to compute options' expected utilities. Evidential probabilities are mathematical representations of empirical relations among doxastic attitudes. Ideal agents exercise reliable forms of reflection. Although humans do not have access to all their degrees of belief, ideal agents know their minds thoroughly. They know whether they believe one proposition twice as strongly as another proposition. An ideal agent has access to her propositional attitudes, including the doxastic attitudes her degrees of belief represent. The mental states causing reflection's reports have the relations that reflection reports. That an ideal agent reports believing a disjunction twice as strongly as each disjunct is good evidence that her mental states are as she reports. An ideal agent has access to evidential probabilities relative to her evidence because she knows the evidence she possesses and knows any degrees of belief it prompts.

In special cases, people approximate rational ideal agents. A person who has the same degree of belief that the first card drawn at random from a standard deck will be a heart as that it will be a diamond typically has a degree of belief that it will be a red card twice as great as that it will be a heart and knows this. Such approximations justify applications of a model for ideal agents to cases in the real world.

If an agent's evidence settles a precise rational degree of belief in some cases, then two ideal agents with the same evidence assign the same evidential probabilities in those cases. If in some cases evidence allows agents to exercise epistemic taste in forming degrees of belief, then two agents with the same evidence may assign different degrees of belief in those cases. Theorists who allow for such differences call evidential probability *subjective probability*. Settling whether rationality gives agents some latitude in using evidence to assign degrees of belief is not important for this book's project. We may assume that rational degrees of belief obey all applicable constraints without fully specifying the constraints.

A general version of the expected-utility principle uses *conditional probabilities*. It considers the probability that an option will have a certain outcome given the option's realization, and it divides possible outcomes using a partition of possible states of the world that settle the option's outcome. Because an option may influence a state's probability, the formula for expected utility uses the state's probability given the option. The standard conditional probability of a state s given an option o, $P(s \mid o)$, equals $P(s \ \& \ o)/P(o)$. This conditional probability registers correlation rather than causation between o and s. Using it to compute expected utility yields evidential decision theory. Worlds, namely, complete states of affairs, may serve as states in calculations of an option's expected utility. An option's world w includes the option's realization (assuming that it is not a matter of indifference). According to the standard conditional probability that evidential decision theory uses, $P(w \mid o)$ equals $P(w \ \& \ o)/P(o)$, and because w includes o, $P(w \ \& \ o) = P(w)$, so $P(w \mid o) = P(w)/P(o)$.

In the formula for expected utility, causal decision theory replaces standard conditional probabilities with *causal conditional-probabilities* so that expected utilities fit their role in rational choice. It replaces $P(s \mid o)$ with a causal conditional-probability $P(s$ given $o)$ that is sensitive to causal relations and insensitive to merely evidential relations between o and s. In many cases, $P(s$ given $o)$ equals the probability of the *subjunctive conditional* that if o were realized, then s would obtain, that is, $P(o \ \square\!\!\rightarrow s)$. The conditional's truth and hence its probability depends on o's effect on s rather than its correlation with s. Gibbard and Harper ([1978] 1981) take the causal conditional-probability as the probability of the subjunctive conditional. They treat decision problems in which each option o has a unique nearest option-world and take the conditional $o \ \square\!\!\rightarrow s$ as true just in case s is true in that world.[6] Because this book similarly treats decision problems meeting this

---

[6] A general analysis of subjunctive conditionals that handles an antecedent that does not generate a unique nearest antecedent-world may say that $A \ \square\!\!\rightarrow C$ holds if and only if either A is impossible or some A-world exists such that C holds in it and in all A-worlds as close or closer than it; that is, some $(A \ \& \ C)$-world is closer than any $(A \ \& \ \sim C)$-world. The analysis may acknowledge that relevant similarities between worlds depend on context, including the conditional itself.

The section's partial analysis suffices for the cases the book treats. In these cases, the box-arrow sentential connective expresses a Stalnaker conditional with the semantics Stalnaker ([1968] 1981) introduces. Specifying a general similarity relation among worlds, or a general selection function from a world and antecedent to a world, is not necessary. Examples rely on simple features of the similarity relation, such as heavy weighting of similarity concerning causal laws.

idealization, it adopts Gibbard and Harper's interpretation of causal conditional-probabilities. For generality, Joyce (1999) takes the causal conditional-probability $P(s$ given $o)$ as the probability image of s given o, and Weirich (2001) takes it as the probability of s if o were realized, a quantity implicitly defined by a theory of causal conditional-probability. Section 6.5 also explicates this quantity.

2.2   **Utility**

Utility is a technical term that the literature introduces various ways. One definition takes utility to be constructed from preferences. To strengthen normative principles of utility, I adopt an alternative account. Utility, as I characterize it, arises from strength of desire. A *degree of desire* indicates the strength of a desire that holds on balance and with all things considered. The measure for degree of desire is traditionally an interval scale. However, common principles of utility neither require nor object to using indifference as a zero point for utility. Jeffrey ([1965] 1990: 82), for example, uses indifference (taken as the desirability of a tautology) as a zero point.

The literature on subjunctive conditionals, or counterfactuals, offers many accounts of their semantics besides the standard similarity account. For example, Gillies (2007) treats counterfactuals as strict conditionals, that is, material conditionals governed by a necessity operator, with a context-sensitive domain of possible worlds. His account accommodates judgments about the consistency and inconsistency of sequences of counterfactuals in conversation. For a single counterfactual that may express the outcome of an option's adoption, his account agrees with the standard similarity semantics. To make Lewis's (1973) analysis of subjunctive conditionals suitable for science, Leitgeb (2012) proposes this modification: $A \:\square\!\!\rightarrow C$ is true in a world w if and only if the conditional chance of C given A at w is very high. Fine (2012) constructs a semantics for counterfactuals that uses states, including world-states, instead of only possible worlds. The semantics' richer resources resolve puzzles about counterfactuals that have disjunctions as antecedents.

Indicative conditionals have a divergent analysis according to some views. Lewis (1976) takes an indicative conditional to be a material conditional and to be assertible when the probability of its consequent given its antecedent is high. Kratzer (2012: chap. 4) claims that an *if*-clause restricts an operator, such as a quantifier, so that *if . . . then* is not a two-place propositional operator, and that the restrictions *if*-clauses create depend on their positions in a conversation. She offers a unified account of both indicative and subjunctive conditionals. Adopting Katzer's thesis that *if*-clauses are domain restrictors, Egré and Cozic (2010) argue that probabilities of simple indicative conditionals with factual antecedents and consequents equal conditional probabilities. Rothschild (2013) shows how to support the view that indicative conditionals express propositions despite puzzles concerning their probabilities. Treating an indicative conditional as a strict conditional, he takes it to be true if and only if the corresponding material conditional is true in every world compatible with the relevant source of knowledge.

I do not propose a full account of conditional sentences and their use in conversations because I need only a simplified account of conditional sentences to introduce tools for evaluation of the options in a decision problem. I use subjunctive conditionals to introduce an option's outcome, future, region of influence, consequences, and types of conditional probability and conditional utility. A full account of subjunctive conditionals may refine utility analysis but does not alter its main features.

A utility scale that uses indifference as a zero point is a ratio scale. Some desire serves as a unit, and other desires have degrees according to their strengths in comparison with the unit desire.

Taking utility to arise from strength of desire makes an agent's utility assignment to an option, or to an option's outcome, express a propositional attitude because, according to a common usage I adopt, desire is a propositional attitude. That the objects of utility are propositions, as are the objects of probability, simplifies decision principles that attach both probabilities and utilities to an option's possible outcomes. The probabilities and utilities alike attach to propositions representing the possible outcomes.

An agent often has access to his degrees of desire. Also, his preferences typically reveal their comparisons. If he prefers an apple to a banana and lacks both, then usually he wants an apple more than he wants a banana. If he is indifferent between getting a banana and getting an apple provided that a coin toss yields heads, then typically he wants an apple twice as much as a banana. Because degrees of desire are defined implicitly by their role in an agent's mental life and by their causes and effects, their conformity to utility laws is a normative requirement rather than a definitional truth.

As I define it, *utility* is a rational ideal agent's degree of desire. An ideal agent's degrees of desire, if rational, obey the principle of agreement with preferences: if an agent prefers x to y and lacks both x and y, then she wants x more strongly than she wants y. Rational ideal agents have degrees of desire that conform with standard principles of utility, including, besides the principle of agreement with preferences, the principle that an option's utility equals its expected utility.

Degree of desire is manifest in choice. A person who trades a bagel and $1 for a doughnut shows that she prefers a doughnut to a bagel and so has a greater degree of desire for a doughnut than for a bagel, assuming that her preference rests on her degrees of desire. Utility does not have a behavioral definition, but its introduction recognizes that behavior is evidence of utility assignments.

Utility analysis uses the normative structure of strengths of desire. The structure may apply to values rather than to desires, so some features of desire do not influence utility analysis. This chapter specifies important features of desire (as understood here) without expounding a full account of desire or a full account of the closely related phenomenon of preference.

Desire has multiple senses. It may be an emotion, or it may be an attitude. Desiring that a proposition hold is an attitude equivalent to wanting the proposition to hold. It need not come with a feeling or emotion. The host of a party, resigned to cleaning up, may want to, or desire to, wash the dishes without experiencing an emotion. Of course, emotion may stimulate desires and preferences. Fear may create a desire for flight.

Parfit (1984) and Schroeder (2004) maintain that desire (in particular, intrinsic desire) is a propositional attitude held with various strengths. Parfit (1984: part 2) examines the present-aim theory of rationality, according to one version of which, what each of us has most reason to do is whatever would best fulfill her present desires. The theory treats desires that a proposition hold, takes wanting as synonymous with desiring, weighs desires according to their strengths, and sums fulfillment of underived desires to evaluate options. Although Parfit extends the scope of desires to include intentions and wishes concerning the past, I assume desire's narrower, traditional scope.

Schroeder (2004: especially chap. 5) explores desire's connections to motivation, pleasure, and reward, taking its connection to reward as primary. Although everyday usage may restrict desire to a passionate state and philosophical usage may generalize it to a pro-attitude, Schroeder takes desire as an intermediate mental state. His account targets intrinsic desire that a proposition obtain, allows for various strengths of desire, and distinguishes desire that a proposition obtain from aversion toward the proposition's not obtaining. It treats desiring as a natural psychological kind of state that we also express as wanting or wishing. Although this chapter agrees with Schroeder's account of desire in taking desiring and wanting as synonymous, it distinguishes desiring from wishing. Desires, but not all wishes, are reasons to act. A surfer may wish she had been born in Hawaii without desiring that birthplace. Knowing that she cannot alter the past, she lacks reasons to act to change her birthplace. Also, because of a desire's role in motivation, a desire fades when fulfilled. A worker who likes a job she has may want to keep it but does not want to win it because she already has won it in competition with other applicants.

Desiring an event's occurrence typically grounds preferring its occurrence to its nonoccurrence, and desiring one event more than another typically prompts a preference for the first event's occurrence. This chapter takes preference to be a mental state closely related to desire. As Binmore (2007: 6–10, 19–20) explains, an alternative theory of revealed preference uses choices to define preferences. It does not use mental states to explain behavior and evaluates choices only for consistency. Hausman (2012: chap. 3) reviews the case against revealed preference theory's definition of preferences using choices. He argues that although economists sometimes say that they define preferences using choices, in practice they use choices to infer, not define, preferences.

Hausman (2012: chap. 4) offers an account of preference as understood in economic practice (rather than as understood in economic theory). According to it, preference is a total comparative evaluation. It compares two goods or two propositions expressing consumption of each good. It settles choices without being defined by choices. According to Hausman's (2012: 43) consequentialism, preferences among outcomes and beliefs about outcomes settle final preferences among options and then a choice. Hausman takes overall but restricted evaluation as the ordinary concept of preference but recognizes additional senses of preference, including total evaluative comparison.

As Hausman does, this chapter takes preference as all things considered, and as comparing two propositions and settling choices without being defined by choices. Unlike Hausman, it takes as the ordinary concept preference with respect to considerations, and lets context fill in the relevant considerations. The senses of preference that Hausman describes come from comparison with respect to considerations by adjusting considerations. For example, enjoying more is just comparison with respect to enjoyment alone. Also, although Hausman takes preference to be reflective and evaluative, this chapter does not define preference to include reflection and evaluation. Basic preferences are not comparative evaluations reflecting on criteria an agent endorses. Preferring one proposition's holding to a second proposition's holding is just wanting the first's truth more than the second's truth. It is a comparison but not necessarily a reflective or evaluative comparison.

In decision theory, a typical representation theorem establishes that if preferences meet certain structural and normative conditions, a probability function (obeying the laws of probability) and a utility function (unique given a unit and zero point) exist such that according to the probability and utility functions, a preference holds between two propositions if and only if the first has greater expected utility than the second, and indifference holds between the two propositions if and only if they have the same expected utility. Savage ([1954] 1972) proves a representation theorem of this sort.

Suppose that a customer at a sandwich shop prefers a ham sandwich to a cheese sandwich and is indifferent between getting a ham sandwich if a coin toss yields heads and getting a ham sandwich if the coin toss yields tails. Then, a probability function representing these preferences assigns the same probability to heads and to tails and so gives each a probability of ½. A utility function that uses a ham sandwich as the unit and a cheese sandwich as the zero point assigns utility ½ to both gambles involving the coin toss.

Representation theorems show that under certain conditions an agent's preferences reveal her utility assignment. The theorems present conditions under which a utility function exists that represents preferences as maximizing expected utility and is the only such function using a specified scale for utility. Suppose that one defines a proposition's utility as this function's value for the proposition. Then, the proposition's utility is relative to a set of preferences. Using a representation theorem to ground a definition of utilities makes their definition change each time a new preference joins the set represented. According to the definition, utility assignments are relative to a set of preferences satisfying the conditions (normally an agent's entire set of preferences at a time). Changing the set of preferences changes the set to which the utility assignments are relative, even if it does not change the method of constructing utility assignments from preferences. A set of preferences grounds utility assignments, and the grounds change when the set of preferences changes, even if the utility assignments over the objects of old preferences do not change. For comparison, suppose that an operational definition of temperature makes it relative to a thermometer. Then, changing the thermometer changes the definition of a liquid's temperature even if by chance the original and the new thermometer yield the same reading. Similarly, any change in the set of preferences changes the definition of a proposition's utility if it is defined using a set of preferences.

Using a representation theorem to ground a definition of utility works only for agents whose preferences satisfy the standard preference axioms. As Kahneman and Tversky (1979) show, humans generally do not satisfy the normative axioms, let alone the structural axioms. Moreover, if an agent at some time has preferences satisfying the axioms, the agent's forming a new preference, if it makes the agent's set of preferences violate the constraints, voids the old assignment of utilities, even if the old preferences do not change, assuming that the agent's utility assignment at a time rests on the agent's entire set of preferences at the time. If an agent violates the preference axioms in any way, say, by having some cyclical preferences, then, strictly speaking, according to a constructivist definition, the agent has no probability and utility assignments. The agent assigns no probability to heads on a coin toss and no utility to gaining a dollar. Constructivism may find ways of working around these problems, but strengths of belief and desire easily accommodate shortcomings; imperfect agents may have degrees of belief and degrees of desire.[7]

---

[7] A constructivist may hold that an agent who violates the preference axioms has the probability and utility assignments that represent the agent's preferences after minimally correcting them so that they satisfy the axioms (assuming that a unique minimal correction exists). Such probability and utility assignments only approximately represent the agent's preferences.

The chapter's realist view of probability and utility gives the representation theorems a normative role and a role in measurement. The existence part of a representation theorem shows satisfaction of a necessary condition for preferences agreeing with expected utilities, namely, the existence of probability and utility functions according to which preferences agree with expected utilities. Meeting the theorem's axioms of preference ensures meeting this necessary condition for satisfaction of a norm of preference. The theorem's uniqueness part grounds measurement of probabilities and utilities by showing that only scale changes distinguish two pairs of probability and utility functions that yield expected utilities agreeing with preferences. Structural assumptions about preferences establish this condition on using preferences to measure probabilities and utilities. A constructive proof of a representation theorem shows how to measure probabilities and utilities. Although revealed preferences do not define utility as this chapter characterizes it, because a rational ideal agent's revealed preferences agree with expected utilities, the unique function that represents the agent as maximizing expected utility identifies the agent's utility function.

Dreier (1996) adopts a constructivist rather than a realist interpretation of utility because he holds that whereas preferences are open to direct introspection, strengths of preference are not. He observes that a typical person who prefers a pomegranate to an apple does not know how much more than an apple she prefers a pomegranate (taking the apple as a unit). However, the imprecision of our desires offers an alternative explanation of our not having a ready answer to some questions about the strengths of our desires. Just as one cannot say how strongly one believes that it will rain tomorrow, one cannot say how strongly one desires a pomegranate because the attitude is imprecise. I put aside such cases by assuming precise probabilities and utilities and by assuming ideal agents with reliable introspective abilities not just concerning their preferences but also concerning their strengths of belief and desire. In any case, even if strengths of desire are not open to introspection, a constructivist definition of them is not the only possibility. One may treat them as theoretical entities, in some cases inferable from preferences, as psychologists treat unconscious desires. Even if observations do not define the theoretical entities, observations may permit inferring their existence. For many theoretical entities, such as temperature, their characterization, or implicit definition, using a theory is more fruitful than their definition using observations. Decision theory's principles of evaluation may use such theoretical entities even if its choice procedures use only accessible states of mind.

Because being an ideal agent entails having access to one's mental states, ideal agents know whether they desire one proposition's realization twice as much as another proposition's realization. Their knowledge of such quantitative relations is as reliable as their knowledge of their preferences. Utility assignments may represent such quantitative relations as well as comparative relations such as preference. A numerical representation of desire decides which empirical relations to represent and how to represent them. Relations among numbers represent empirical relations because of a homomorphism between a numerical relational structure and an empirical relational structure. The numerical relations can represent quantitative relations of desire (within a margin of error) that reflection reports. Some branches of psychology mistrust these reports. They note that a person may not know his own desires. To be sure, a person is not 100 percent reliable, but his reports are good evidence of his desires. A person's choices are similarly not 100 percent reliable evidence of his preferences because of the possibility of slips when distracted and the like. Yet, choices are good evidence of preferences in many cases, and reports are good evidence of quantitative relations of desire in many cases. An ideal agent's choices and reports are, respectively, reliable indicators of preferences and quantitative relations of desire. In special cases, people approximate ideal agents. A person who has the same degree of desire for gaining a dollar if a die roll yields an ace as for gaining a dollar if the roll yields a deuce typically has twice the degree of desire for gaining a dollar if the roll yields an ace or a deuce and knows this.

Suppose that John sees a film about Florence and then wants more strongly to visit the city. Consequently, among vacation destinations, he prefers Florence and decides to go there. This explanation uses strength of desire as a cause of preferences among options and then a choice. Defining utility using strength of desire gives utility this explanatory role. The definition using preferences allows preferences to explain utilities but does not allow utilities to explain preferences and choices.

Decision theory gains by representing propositional attitudes such as desire to a certain degree in addition to representing relations such as preference. Degrees of belief and degrees of desire explain preferences. They even explain incoherent preferences that cannot be represented as maximizing expected utility, as in cases that Kahneman and Tversky (1979) and Baron (2008) report. Incoherent preferences support the existence of degrees of desire not reducible to properties of coherent preferences. Realistic decision theory treats mental states besides preferences. It has norms for degrees of belief and degrees of desire, not just for preferences.

Behavioristic theories of probability and utility treat preference's structure and preference's effects on choice but often gloss over preference's origins. An account of preferences' origins describes evidence of preferences besides effects. A full theory of preference does not define preference in terms of its effects, ignoring the causes of preferences. An account of preferences' origins distinguishes basic and derived preferences. It examines the causes of basic preferences, which may be experiences such as pleasure and discovery of other people's preferences. Besides good marketing, a person's social role and cultural identity influence basic preferences. Humans have innate desires such as desires for sleep, food, sex, and a moderate ambient temperature. Unusual causes are possible, of course. A preference may arise from a blow to the head, drugs, and hypnosis. Although inferring preferences from the common causes of preferences is not infallible, behavioral science should not adopt higher standards for conclusions about preferences than it adopts for conclusions about other mental phenomena.

A person's information is evidence of her degrees of belief. If she knows that a coin is fair, then most likely her degree of belief that it lands heads when tossed is 50 percent. Similarly, a person's not having slept for 24 hours is a sign that she wants sleep. In normal circumstances, a person's knowing that a bet on a fair coin pays \$4 if heads and nothing otherwise is evidence that she wants the bet twice as much as she wants \$1. An agent's self-reflection is also a source of evidence about her degrees of desire.

Principles of rational decision use an agent's psychological states, especially the agent's belief-states and desire-states. A normative principle may use states whose identification relies on the states' causal roles. A belief-state's identification may use the state's causal role. The belief-state so identified may figure in a normative principle of consistency and a normative principle of decision. A desire-state's identification may likewise use the state's causal role, including its causes. It need not have a behavioral definition that uses only the state's effects.

An introduction of an option's utility tailors it for the decision principle of utility maximization. Because an option's utility evaluates an option's choice-worthiness, it processes all considerations relevant to an option's choiceworthiness. To be safe, it evaluates an option's outcome taken as a possible world and given incomplete information surveys the worlds that might be the option's outcome. A maximal consistent proposition represents a possible world. Simplicity trims a possible world so that it covers only features that the agent cares about. It is a proposition that says for each feature an agent cares about whether that feature obtains. For economy, the features may be just the objects of basic intrinsic attitudes. Taking worlds as propositions, probabilities and utilities attach to worlds. An option's utility attaches to a proposition representing the option. If o is an option, the option's utility $U(o)$ evaluates o's *outcome* taken as a possible world. Let $O[o]$ be a sentential expression of o's outcome's obtaining, rather than a non-sentential name of o's outcome. Then, taking U to evaluate comprehensively, $U(o) = U(O[o])$. This equation holds even given ignorance of o's outcome.

To illustrate, suppose that if one bets, then one will win, although one does not know this. Because one does not know that winning is betting's outcome, $U(\text{bet}) \neq U(\text{winning})$. However, $O[\text{bet}]$ is that betting's outcome obtains; it does not express that one wins. Hence, $U(\text{bet}) = U(O[\text{bet}])$.

Subjunctive supposition of an option's realization identifies a nearest option-world using a similarity relation among possible worlds that gives priority to causal relations. If an option's realization would cause an event, then typically that event is part of the nearest option-world.

An agent may not know the world that would be realized if she were to realize an option. She may be ignorant of the option's world because she does not know which world results from the option's realization. Which possible world is the option's world may be unknown, partly because the causal relations settling the nearest option-world are unknown. Various possible worlds may be candidates for the option's world. A probability distribution may apply to them. Whether a possible world might obtain if an option were realized is an epistemic question. The word "might" expresses epistemic possibility.

Although an option's world is its outcome taken comprehensively, utility assignments use different propositions to represent realization of an option's world and realization of its outcome. A maximal consistent proposition represents a world's realization. The proposition that the option's world obtains represents realization of an option's outcome. It is not a maximal proposition. Ignorance of the option's world may make the utilities of the two propositions differ. Given incomplete information about a lottery ticket's prospects, an ideal agent does not know the utility she would assign to owning the ticket if she had full information. She would assign a high utility if she were to know that the ticket will win and a low utility if she were to know that it will lose. However, she does not know whether it will win or lose. In such a case, an agent may not know an option o's world $W[o]$ and so may not know $U(W[o])$ even though U is her own utility assignment. Nonetheless, the agent knows the proposition that the option's world obtains and so knows $U(O[o])$. $U(O[o])$ treats $O[o]$ as a lottery among the various worlds that might be o's world. It is a probability-weighted average of their utilities rather than the utility of o's world. It is possible that $U(O[o]) \neq U(W[o])$.

According to the expected-utility principle, an option's utility equals the option's expected utility, a probability-weighted average of the utilities of its possible outcomes. Some versions of the expected-utility principle replace the utility of a possible outcome of an option o with the utility of o *conditional* on realization of a state s from a partition of states. The utility of o given s, $U(o$ given $s)$, is not the utility of the conjunction of o and s because the expected-utility principle supposes o and supposes s in different ways, whereas $U(o \& s)$ supposes their conjunction just one way. In routine cases, the principle supposes o subjunctively, or causally, whereas it supposes s indicatively, or evidentially. Section 2.5 treats the general case. Weirich (2001: sec. 4.2.2) implicitly defines conditional utility using a utility theory grounding the expected-utility principle.[8]

2.3   Expected utility's computation

The expected-utility principle evaluates an option in a decision problem using the probabilities and utilities of the option's possible outcomes. The principle's normative status depends on the interpretation of the probabilities and utilities. If probability and utility are constructively defined to make the option's utility equal its expected utility, then the principle has no normative force; it is merely a conventional constraint on representations of preferences using probabilities and utilities. Any normativity resides in the principles of preference that ensure the possibility of representing preferences "as if" they were maximizing expected utility. This chapter's realist interpretation of probability and utility yields a stronger version of the expected-utility principle that grounds a requirement that preferences not only be "as if" maximizing expected utility but also genuinely maximize expected utility. The strengthened expected-utility principle guides the formation of preferences and does not just create a conventional representation of preferences already formed.

Savage ([1954] 1972) shows that for an agent whose preferences meet his axioms, preferences are "as if" maximizing expected utility. Some axioms of preference, such as transitivity, are normative constraints, but the expected-utility principle itself is just a conventional constraint on representations of preferences that use probabilities and utilities. This chapter's version of the

---

[8] Bradley (1999) rejects the equation $U(o$ given $s) = U(o \& s)$. He takes conditional utility as marginal utility, as Chapter 3 does for intrinsic utility.

expected-utility principle makes meeting the principle a normative constraint in addition to the normative preference axioms. The chapter thereby strengthens Savage's decision theory. A theory of rationality may adopt his normative axioms of preference (using his representation theorem to ground a method of measurement rather than a definition) and then adopt as an additional norm a realist version of the expected-utility principle.

Textbook decision problems, such as decisions about bets on a coin toss, assume that the outcomes mentioned, such as monetary gains and losses, are separable from other features of the worlds that the options realize. However, this separability does not hold in all decision problems. Suppose that a traveler has an opportunity to gain a dollar. She may decline the dollar because it puts her over the threshold for custom declarations of gains while abroad. Being tired, she would rather forgo the dollar than fill out the declaration. An increase in wealth does not produce an increase in comprehensive utility. Monetary outcomes are not separable from non-monetary outcomes in this case.

Because in many cases an option's simplified outcome is not separable from other features of the option's world, world-Bayesianism uses worlds as outcomes. According to world-Bayesianism, which Sobel (1994: chap. 1) presents, an option's possible outcomes are the possible worlds that might be realized if the option were realized. Only a finite number of worlds are deliberationally significant given that an agent's desires distinguish only a finite number of metaphysical possible worlds. Thus, an option's utility, its *expected utility*, is a probability-weighted sum of the utilities of possible worlds suitably interpreted. The formula for an option's expected utility is $\sum_i P(w_i \text{ given } o) U(w_i)$, where w_i ranges over the possible worlds or, more precisely, the propositions that represent the possible worlds. The summation may, however, drop worlds in which o is unrealized because their probabilities conditional on o are zero. Hence, the formula may take w_i to range over the possible worlds that might be the option's world. Worlds excluded, conditional on o, have probability zero.[9]

So that an ideal agent has access to an option's utility despite incomplete empirical information, the principle of utility maximization takes an option's utility to equal not the option's informed utility but the expected

---

[9] An epistemically possible world may have probability zero in some probability models, but the formula's including such a world does no harm. In fact, if a proposition with probability zero is epistemically possible, then deliberations that ignore it may go wrong. To dodge difficulties such propositions cause, this book treats only decision problems in which deliberations may use a finite Archimedean probability space. In such decision problems, deliberations may safely ignore propositions with probability zero.

value of the option's informed utility. That is, an option o's utility equals o's expected utility $EU(o)$ taken as the expected utility of o's world rather than the utility of o's world. Consequently, $U(o) = EU(o) = \sum_i P(w_i$ given $o)$ $U(w_i)$, where w_i ranges over worlds that might be o's world. $U(o)$ is sensitive to information, although $U(w_i)$ is not because w_i is a maximal proposition. $U(o) = EU(o)$ expresses a normative principle for degrees of desire, to which a rational ideal agent conforms. The principle assumes that the requisite probabilities and utilities exist and are known to the agent.

The formula for expected utility, and so for an option's utility in the case of an ideal agent, belongs to causal decision theory because it uses the probability of a world given subjunctive supposition of the option's realization. It uses the probability that the world would be realized if the option were realized. This type of supposition gives priority to causal relations in settling the nearest option-world and so, agreeing with causal decision theory, makes an option's consequences settle the option's utility.

Specifying subjunctive supposition of the option yields an accurate version of the decision principle of expected-utility maximization, given the chapter's idealizations. The main alternative, taking the utility of an option to involve an indicative supposition of the option, yields an incorrect version of the principle because evidential relations direct indicative supposition. Because subjunctive (causal) and indicative (evidential) supposition use different measures of distance between possible worlds, an option o's outcome if o *were* realized may differ from o's outcome if o *is* realized. The outcome if an option is performed may be attractive because of the evidence the option provides rather than because of its promise of good consequences.[10]

Maximizing evidential attractiveness, instead of prospects of good consequences, leads to bad decisions in cases such as Newcomb's problem. In Newcomb's problem, an agent has a choice between taking an opaque box that may contain money or taking the opaque box together with a transparent box containing \$1,000. A good predictor, before the agent's choice, places \$1,000,000 in the opaque box just in case she predicts that the agent will take only the opaque box. The agent's choice does not affect any box's contents. Taking just the opaque box is evidence that the box contains \$1,000,000, although it does not affect the box's contents. It has maximum evidential attractiveness. However, the agent is \$1,000 ahead taking both

---

[10] Causal decision theory uses indicative supposition of an option to check whether the option is self-ratifying, that is, to compute its utility given that it is realized and to check whether under that supposition its utility matches or exceeds every other option's utility.

boxes no matter what the opaque box contains. Taking both boxes maximizes the prospects of good consequences. The rational choice, assuming causal decision theory, is to take both boxes.

Although the formula for expected utility registers an option's effect on the probabilities of worlds, it does not divide an option's world into the option's consequences and other events. The formula makes an option's consequences control the option's expected utility, but the formula attends to an option's world and not exclusively to an option's consequences. World-Bayesianism evaluates an option by evaluating the worlds that might obtain if the option were realized. The worlds include the past and elements of the future that would occur whether or not the option were realized. However, causal decision theory maintains that an option's evaluation may attend to only the option's consequences, and later chapters trim other events from an option's evaluation.

2.4   Outcomes

The previous section's expected-utility principle takes the outcome of an agent's option to include every event the agent cares about that occurs given the option's realization. Taking outcomes comprehensively in this way makes the individuation of outcomes fine. A single, small event may distinguish two outcomes. Does a fine individuation of outcomes complicate rather than simplify choices, and so thwart the book's project? The project is to evaluate options as efficiently as possible without losing accuracy. A fine individuation of outcomes is necessary for accuracy; a coarse individuation omits considerations that matter to an agent and to an option's evaluation. The expected-utility principle's plausibility depends on a fine individuation of outcomes.

Some theorists fear that a comprehensive account of an option's outcome makes the expected-utility principle trivial. It permits rationalizing any preference whatsoever. Whenever an agent appears to violate the principle, one may add to outcomes features that preserve compliance with the principle. If a shopper buys the more expensive of two identical laundry detergents, one may preserve maximization of expected utility by adding to the purchase's outcome possession of the detergent's distinctive packaging.

Fine individuation of outcomes, the objection claims, also trivializes constraints on preferences, such as transitivity, because it may transform any apparent violation of a constraint into a compliant case. Given an agent's apparent preferences for *A* over *B*, *B* over *C*, and *C* over *A*, one may distinguish A compared with B from A compared with C to break the cycle.

As Dreier (1996) notes, however, a fine individuation of outcomes does not plausibly eliminate every violation of transitivity. In the example, if A, whether compared with B or with C, is the same in every feature that matters to the agent, then nothing justifies distinguishing A compared with B from A compared with C to eliminate the cycle. Dreier defends, in an agent's decision problem, an individuation of outcomes as fine as necessary to include all that matters to the agent (even if the individuation makes preferences between some outcomes impractical because the agent cannot ever be in a position to choose between them). This response to the fear of trivialization supports a fine individuation of outcomes. An agent's desires limit individuation of outcomes to preserve the substance of norms of preference and utility.

Savage's ([1954] 1972: sec. 2.7) *sure-thing principle* (a preference axiom for his representation theorem) asserts, for a preference order of options represented by their consequences with respect to a partition of states, the separability of the preference order of options' consequences given a state. According to the principle, the order of two options agrees with the order of their consequences in a state s, given that they have common consequences in s's complement. This normative principle applies pair wise to options to order them given s.

If outcomes are coarsely individuated, then objections to the principle arise. For example, suppose that outcomes specify only monetary consequences although an agent also cares about risk. Not all agents have an aversion to risk, but imagine an agent who is averse to risk. Two options both yield \$10 in a certain state and differ only in the state's complement. Given its complement, one option yields \$10 while the other yields \$0 or \$20 with equal probability. The agent may prefer the first option because of aversion to risk. Next, suppose that the options yield \$0 in the state but have the original payoffs in the state's complement. Given the change, if the state's complement is unlikely, the agent may prefer the second option to the first option because they differ little in risk and the second offers a chance for a larger prize. This preference reversal, although justified by aversion to risk, is contrary to the sure-thing principle applied with respect to monetary consequences only.

However, the sure-thing principle is plausible if consequences are comprehensive and not just monetary. As Weirich (1986) argues, Allais's Paradox, which challenges the sure-thing principle, arises from taking consequences narrowly so that they include just amounts of money.

Section 2.7 argues thoroughly for a principle of separation similar to Savage's sure-thing principle. Its principle of separation is immune to Allais's objection because it finely individuates outcomes.[11]

Broome (1991: 115–17) observes that a fine individuation of outcomes threatens some structural axioms of preference on which representation theorems rely. The axioms require that the same outcomes occur in many contexts and so hold generally only given a coarse individuation of outcomes. A constructivist theory that uses the representation theorems to define utility may therefore reject a fine individuation of outcomes. However, a realist theory that uses the theorems only to introduce methods of measuring utility may finely individuate outcomes and restrict the methods of measuring utility to cases complying with the structural axioms. Putting aside pressure to make compromises for the sake of measurability, this is the better course if the normative axioms are plausible only given a fine individuation of outcomes.

Chapter 3 analyzes utilities using basic intrinsic desires and aversions. It takes finely individuated outcomes as realizations of sets of basic intrinsic attitudes, including any basic intrinsic attitude toward an arrangement of realizations of other basic intrinsic attitudes. Accurate principles of utility analysis take outcomes comprehensively and so individuate them finely.

2.5   Rational decisions

A standard decision principle for an ideal agent asserts that an option's realization is rational if and only if the option maximizes utility; that is, its utility is at least as great as any other option's utility. In a rational ideal agent, an option's utility equals its expected utility, so the principle also supports expected-utility maximization. This section reviews the principle of utility maximization to elucidate utility by explicating its normative role.

Rational choice at a moment depends on beliefs and desires at the moment, although a rational person cares now about satisfaction of future desires. The principle of utility maximization rests on an idealization. It assumes that present beliefs and desires are rational and so applies to agents without present, imprudent desires. The idealization puts aside cases in which irrational input for utility maximization justifies compensatory departures from utility maximization.

---

[11] The principle that Section 2.7 supports does not endorse all assumptions of Savage's representation theorem; it does not endorse the assumption that for anything an agent cares about, some option yields it in every state.

To be rational in all respects, a decision must not only maximize utility but also win adoption because it maximizes utility. A person who chooses a $5 bill over a $1 bill makes the right choice. However, if he chooses by whim, he chooses correctly for the wrong reason. A fully rational agent meets high standards and performs the right act for the right reason. Advancing the principle of utility maximization for a fully rational agent ensures not only that the input for the principle is rational but also that in cases of compliance with the principle the right reasons generate compliance.

The principle of utility maximization, stated as a necessary and sufficient condition of rationality, assumes not only an ideal agent who is fully rational except perhaps in adopting an option to resolve the current decision problem but also an ideal decision problem with an option of stable maximum utility. Rationality requires that a rational ideal agent in an ideal decision problem realize an option that maximizes utility, that is, expected utility.

For simplicity, the principle's idealizations are stronger than necessary. They put aside cases in which probabilities or utilities, despite obeying standard laws and so qualifying as probabilities and utilities according to minimal standards, are not fully rational degrees of belief and fully rational degrees of desire. Taking probabilities and utilities, respectively, as degrees of belief and degrees of desire in a fully rational ideal agent has the same effect. The idealizations also ensure that an agent considers all her options. Her realization of an option may not be rational despite its maximization of utility if she is unaware that it maximizes utility because she does not consider all options. However, to weaken idealizations, the principle may weaken its claim by asserting just that an option is rational to realize if and only if it maximizes utility. By not evaluating an agent's realization of an option, it may dispense with idealizations that ensure consideration of all options.

An agent in a decision problem may perform an act, realize an option, and maximize utility among a subset of options. The options to which the principle of utility maximization applies do not include all options, and options do not include all acts. An agent's types of control of her acts characterize options and the subset of options to which the principle of utility maximization applies. Treatments of decision problems typically count as options only the options in a suitable subset.

The decision principle to maximize utility assumes an account of an agent's options in a decision problem. Ordinary language takes her *options* as free acts she fully controls at the decision problem's time. This account of options adopts the ordinary concept of freedom but does not offer a philosophical analysis of freedom. An agent *fully controls* an act if and only if she directly controls it or its components. An agent *directly controls* an act if and only if she may perform it at will. She fully controls its components if and only if she directly controls each component at the time for it. Hence, she fully controls a walk if she directly controls each step of the walk at the time for the step.

Some options are decisions. An agent decides at will and so directly controls her decisions. Some options are sequences of acts an agent performs at will. An agent performs at will the sequence of steps that constitutes her walk and so fully controls her walk.

The options that utility maximization compares are those in an agent's direct control. Utility maximization compares them with other options in an agent's direct control at the time of the decision problem. Direct control implies immediate control. So utility maximization puts aside an act an agent does not immediately control but performs later because of an act in her immediate control. A homeowner adjusts the thermostat now and so warms the room, but warming the room occurs later and is not in her immediate control. Warming the room is not among the set of options utility maximization targets in the homeowner's current decision problem about the thermostat's setting. (A decision principle may count as options in the agent's direct control some that the agent does not directly control if the difference does not matter in the decision problem at hand.)

Utility maximization, moreover, puts aside all acts that are not options. Consider starting a race. Although it is an act, it is not an option. A starter may perform the act by firing a starting pistol. However, the runners partly control the race's start. So, the starter does not fully control it. It is not the starter's option, but just a consequence of the starter's option to fire the pistol. It has a utility for the starter, but rationality does not apply the standard of utility maximization to it. Rationality just gives firing the pistol a positive evaluation because the option has maximal utility, among options in the starter's direct control, in virtue of having starting the race as a consequence with high utility.

Rationality does not apply the standard of utility maximization to options that an agent does not control directly. If an agent fully but not directly controls an act, then the act is composite. If it were simple and in the agent's full control, it would be in the agent's direct control. To evaluate an act fully but not directly controlled, rationality evaluates the act's components. They are acts that the agent directly controls. If they maximize utility and so are rational, then the composite act is rational. This is a sufficient but not a necessary condition of a composite act's rationality because a rational composite act may have an irrational but insignificant component. For example, a rational walk may have an irrational step that has negligible consequences for the walk. Irrationally stepping in a puddle need not make a walk irrational because the other rational steps may suffice to make the walk rational.

The principle of utility maximization and the derivative principle of expected-utility maximization apply to the set of options in an agent's direct control at the time of the decision problem. They do not apply to the entire set of the agent's options, including options that the agent fully but not directly controls. The options in a decision problem are acts over which the agent has direct control.

Decisions are paradigm examples of acts an agent directly controls, and decision theory attends to their evaluation. The decisions are in the agent's direct control even if executing the decisions and performing the acts that follow from their executions are not. Deliberations may take options as possible decisions rather than as the acts that form the contents of those possible decisions. Standard forms of utility analysis assume that options all have the same time of realization, and the possible decisions that form a decision problem's possible resolutions satisfy this assumption. Taking options as decisions makes options certain to be carried out if adopted and so simplifies decision principles. To follow tradition, this book takes options to be decisions. It evaluates only decisions and treats only decision problems in which the options are decisions.[12]

Treating only the decisions that the agent might reach simplifies many decision problems with options besides decisions. Evaluations of decisions to perform acts may often replace evaluations of the acts themselves. However, for decision problems that require evaluation of acts besides decisions, Weirich (2010a) formulates methods that depend on the type of control the agent has over the acts.

A concrete decision problem has multiple representations. Take a decision about buying a hat. One list of options may present two possible decisions: (1) to buy a hat and (2) to not buy a hat. Another list may present three possible decisions: (1) to buy a red hat, (2) to buy a hat of another color, and (3) to not buy a hat. The lists do not conflict because buying a red hat and buying a hat of another color are ways of buying a hat. Accurate representations of a decision problem may list different options.

Textbook representations simplify by listing only salient options instead of all relevant options. A full list includes every decision an agent might

[12] Hedden (2012) describes the advantages of taking options as decisions.

make to resolve the problem. A solution of a decision problem is invariant with respect to accurate representations of the problem, although a solution may be differently characterized according to the terms of different representations. Standard game matrices do not represent all options, just exclusive and exhaustive pure strategies; they omit mixed strategies. So that a Nash strategy with respect to a matrix is utility maximizing with respect to all strategies, the Nash strategy must be implemented in the utility-maximizing way. An option's utility is the utility of the best way of performing it if there are several ways of performing it. For a basic act, just one way of performing it is in the agent's control.[13]

Some representations of a decision problem ignore the past. Principles of separability show that a representation using options' futures is equivalent to a representation using options' worlds; it yields the same comparisons of options and so the same solutions to decision problems. Principles of separability also support representations that depict only options' regions of influence or only options' consequences. Chapters 4–6 establish these points.

The normative status of the principle to maximize utility depends on the definition of utility, and the normative status of the companion principle to maximize expected utility depends on the definitions of probability and utility. A definition of utility that constructs it from preferences, so that according to the definition preferences maximize utility, makes the principle of utility maximization a definitional truth and not a normative principle. Section 2.2 derives utility from strength of desire to make the principle normative.

Definitions of probability and utility that use preferences entail that preferences agree with expected utilities. The definitions therefore do not ground the normative principle of expected-utility maximization. They support only a weak form of the principle that requires having preferences with an expected-utility representation, so that choices resting on preferences are "as if" maximizing expected utility. In contrast, taking probability and utility as rational degree of belief and rational degree of desire, respectively, grounds the traditional strong form of the normative principle of expected-utility maximization. It demands that choices maximize expected utility according to probabilities and utilities that make an option's expected utility definitionally independent of an agent's preferences.

---

[13] Peterson (2009: chap. 2) discusses representations of decision problems. Samuelson (1997: 30, 48) describes construction of a generic game that conflates some options for an agent into one option and so simplifies representation of the agent's decision problem.

Suppose that for some agent, the only way to have preferences that are "as if" maximizing expected utility is to have preferences that maximize expected utility. For the agent, meeting the weaker norm amounts to meeting the stronger norm, too. Nonetheless, having preferences that are "as if" maximizing expected utility does not in general ensure having preferences that maximize expected utility. If an agent arbitrarily selects a probability function and a utility function and then forms preferences that maximize expected utility according to these arbitrary functions, the preferences meet the weaker norm but, except by incredible chance, fail to meet the stronger norm that obtains expected utilities from the agent's actual probability and utility functions.

Suppose as an empirical generalization that humans have preferences "as if" following expected utilities if and only if their preferences follow expected utilities. The "as if" norm is still weaker than the realist norm, which uses probabilities and utilities not constructed from preferences. For comparison, consider the following norms: (1) drink no beers and (2) drink no more than one beer. The second is weaker, but an alcoholic may meet the second if and only if he meets the first. The relative strength of norms does not depend on empirical relations between their realizations.

Because an ideal rational agent has preferences that maximize expected utility, she also has preferences that are "as if" maximizing expected utility. A probability function and a utility function (given a unit and a zero point) may represent her preferences as maximizing expected utility. Suppose that the structure of her preferences make them the only such functions. Defining her probability and utility functions as the ones that make the representation succeed, although it makes "as if" maximization agree with genuine maximization, has a theoretical cost. Because preferences define probabilities and utilities, probabilities and utilities no longer explain preferences.

The realist view of utility does not undermine the constructivist's argument, using axioms of preference, for having preferences that are "as if" maximizing expected utility. It preserves that argument but notes that the argument does not support the stronger norm of having preferences that maximize expected utility according to probability and utility functions defined independently of preferences. Section 2.7 argues for genuine, as opposed to "as if," maximization of expected utility.

2.6   Partition invariance

Calculation of an option's expected utility uses a partition of possibilities and gives only epistemic possibilities positive weight. World-Bayesianism uses a partition of possible worlds, but sets of possible worlds (besides singletons) may also form partitions. A set of possible worlds is called a *state*, and a proposition represents it. The proposition is true in all and only possible worlds in the set it represents; it is equivalent to a disjunction of the possible worlds in the set. An option's outcome given a state corresponds to a set of possible worlds in which the state and option hold. Which set depends on the forms of supposition of the option and the state. Causal decision theory specifies the forms of supposition in its introduction of the probabilities and utilities that yield the option's expected utility.

Using a partition of worlds to generate a set of possible outcomes ensures that the set neither omits nor double counts relevant considerations. Using partitions of states that represent sets of possible worlds simplifies an option's evaluation by considering whole sets of possible worlds at once instead of considering possible worlds one by one. However, using states to partition possibilities requires changing world-Bayesianism's formula for an option's expected utility. The new formula makes the option's expected utility the same no matter which partition of possibilities the formula uses. This property of the formula is its *partition invariance*. World-Bayesianism is the foundation of decision theory. Forms of Bayesian decision theory using partitions of states work because they are equivalent to world Bayesianism, as Sobel (1994: chap. 9) shows.

The general formula for an option's expected utility uses states instead of worlds and replaces the utility of a world with the utility of an option given that a state obtains. According to it, $EU(o) = \sum_i P(s_i \text{ given } o) U(o \text{ given } s_i)$, where s_i ranges over states in a set partitioning possibilities. The formula yields an option's utility for ideal agents in ideal decision problems. In ideal cases, $U(o) = EU(o)$.

A formula for expected utility that uses states may substitute, for an option's utility given a state, the utility of the option's outcome given the state. In the formula, a state represents a set of possible worlds that may not specify only relevant matters. An option's full outcome given a state may vary with the worlds in the set that the state represents, so the formula typically takes an option's outcome given a state as gains or losses of some type rather than as the option's full outcome. The option's simplified outcome is constant across the worlds in the set. It is the disjunction of the option's full outcomes in those worlds, and its utility is a probability-weighted average of the full outcomes' utilities.

As explained, a possible world is a proposition that fully specifies relevant events. A world in this sense represents an option's comprehensive outcome. This section's general formula for expected utility, $EU(o) = \sum_i P(s_i \text{ given } o)U(o \text{ given } s_i)$, does not replace an option's utility given a state with the utility of the option's outcome given the state. However, the formula may take the outcomes themselves as states. If the outcomes are possible worlds taken as propositions and serve as states, then the formula using states returns world-Bayesianism's formula using worlds. The formula $EU(o) = \sum_i P(w_i \text{ given } o)U(w_i)$ is a special case of the formula using states. In the special case, for each i, s_i is w_i, $P(s_i \text{ given } o)$ equals $P(w_i \text{ given } o)$, and o given s_i is o given w_i. When $P(w_i \text{ given } o)$ is positive, w_i entails o so that $U(o \text{ given } s_i)$ equals $U(w_i)$. Hence, the formula for states, assuming that it may use any partition of states, including the set of worlds, entails the formula for worlds.[14]

Because a calculation of an option's expected utility may use many partitions, using any partition must yield the same quantity if an option's expected utility is to equal the option's utility, in conformity to the expected-utility principle. Weirich (2001: appendix A.2) shows that all partitions yield the same quantity so that an option's expected utility is partition invariant.

The partition-invariant formula for expected utility uses the utility of an option given that a state obtains, $U(o \text{ given } s)$. In a conditional utility, the condition may be any proposition, including one that a subjunctive conditional expresses, in particular, the subjunctive conditional that if the option o were realized, then the state s would obtain, that is, $o \;\square\!\!\rightarrow s$. Under the chapter's assumptions about subjunctive conditionals, $U(o \text{ given } s)$ equals the option's utility given that if the option were performed the state would obtain, that is, $U(o \text{ given } (o \;\square\!\!\rightarrow s))$. The equality of the two conditional utilities explains the type of supposition conditional utility involves. Both utilities equal $U(s)$ if s is a world that entails o's realization. The simplification holds because U is comprehensive and evaluates a proposition's comprehensive outcome.

To illustrate, suppose that the option is accepting a gamble that pays \$1 if a coin toss yields heads, and the state is that the toss yields heads. According to the identity of conditional utilities, the utility of the gamble given heads equals the utility of the gamble given that if it were accepted, then the toss would yield heads.

---

[14] A means of deriving the formula for worlds from the formula for states adopts the identity $U(o \text{ given } s_i) = \sum_{w_j \in s_i} [P(w_j \text{ given } o) / P(s_i \text{ given } o)]U(w_j)$, taking a set of worlds to represent a state. It follows that $\sum_i P(s_i \text{ given } o)U(o \text{ given } s_i) = \sum_i P(s_i \text{ given } o) \sum_{w_j \in s_i} [P(w_j \text{ given } o) / P(s_i \text{ given } o)]U(w_j) = \sum_i \sum_{w_j \in s_i} P(w_j \text{ given } o)U(w_j) = \sum_j P(w_j \text{ given } o)U(w_j) = EU(o)$.

Using $U(o$ given $s)$ rather than $U(o$ given $(o \;\square\!\!\rightarrow s))$ in the formula for an option's expected utility accommodates cases in which $(o \;\square\!\!\rightarrow s)$ lacks a truth-value, and $U(o$ given $(o \;\square\!\!\rightarrow s))$ is therefore inadequate.[15] Although the conditional has a truth-value given the existence of a nearest option-world, using $U(o$ given $s)$ anticipates generalizing the expected-utility principle to dispense with this idealization about option-worlds.

To illustrate the type of supposition of a state s in a conditional utility $U(o$ given $s)$, consider the following case. John is part of a crowd in a lecture hall and considers whether to rise to close a door to keep out noise. He knows that it is unlikely that someone nearer the door will close it. He wants to be helpful, and so he ought to rise to close the door. However, John knows that because he fears attention, he will not rise to close the door.[16]

If the door will be closed (c), then because John will not close it, someone else will close it. So if John were to rise to close it (r), his effort would be in vain. If conditional utility supposes a condition indicatively, $U(r$ given $c)$ is lower than $U(\sim r$ given $c)$. So $EU(r)$ is lower than $EU(\sim r)$. However, $U(r$ given $(r \;\square\!\!\rightarrow c))$ is higher than $U(\sim r$ given $(\sim r \;\square\!\!\rightarrow c))$ because John wants to be helpful, and the first utility does not involve the supposition that John's rising is superfluous. It allows for the influence of John's belief that if he rises, he will be responsible for the door's closing. So calculated using the more complex utilities, $EU(r)$ is greater than $EU(\sim r)$, as it should be.

Using $U(r$ given $c)$ with c supposed indicatively miscalculates $EU(r)$ because the conditional utility ignores the influence of John's rising on the door's closing. Because it assumes the door's closing, it puts aside the possibility that John's rising and subsequent acts cause the door's closing. In general, suppose that an agent desires that an option o cause a state s. This desire does not boost $U(o$ given $s)$, assuming s's indicative supposition, even if the agent believes that o causes s. The indicative supposition that s makes the belief idle. The supposition puts s in the background for entertaining o and so precludes o's causing s. Although $U(o$ given $s)$ is comprehensive and entertains worlds, not just causal consequences, the supposition of s carries

---

[15] Imagine that the conditional lacks a truth-value. Its supposition's taking it to be true may bias the option's evaluation. Consider the conditional that if Verdi and Bizet were compatriots, then they would be Italian. Italy and France tie as the composers' common country, supposing them to be compatriots. So, the conditional lacks a natural truth-value (although some accounts of conditionals by convention make the conditional false). Supposing that the conditional is true arbitrarily breaks the tie and so may bias an option's evaluation that uses the supposition.

[16] John decides irrationally and so is not fully rational, but the case assumes that he is fully rational except in his current decision problem. Consequently, he satisfies idealizations of the principle of expected-utility maximization. The interpretation of conditional utility anticipates generalizing expected-utility maximization to cases in which an agent is not fully rational.

implications about causal relations and so directs evaluation to a world or set
of worlds where o does not cause s.

To obtain a conditional utility that registers the influence of o on s,
instead of having utility conditionalize on s in a way that assumes s is fixed,
this chapter has utility conditionalize on s in a way that is equivalent to
conditionalizing on the conditional that $(o \,\square\!\!\rightarrow s)$. It interprets $U(o$ given $s)$
so that it is equal to $U(o$ given $(o \,\square\!\!\rightarrow s))$. The latter quantity is the utility of
the outcome if the option were realized given that it is the case that the state
would obtain if the option were realized. In the latter quantity, s is supposed
subjunctively as part of a condition supposed indicatively. The change in
type of supposition for s makes the utility sensitive to o's causal influence on
s. The utility rises if the agent believes that o causes s because the supposition
that $(o \,\square\!\!\rightarrow s)$ leaves open whether o causes s. It is unlike the indicative
supposition that s obtains, with its implication in the context of $U(o$ given $s)$
that s obtains independently of the option realized. Altering the type of
supposition of s so that it agrees with the indicative supposition of the
subjunctive conditional accommodates cases in which the option has a
desirable or undesirable influence on the state. It makes expected-utility
analysis count that relevant consideration. In the example, expected-utility
analysis with $U(r$ given $c)$ interpreted so that it equals $U(r$ given $(r \,\square\!\!\rightarrow c))$,
and similarly for other conditional utilities, yields the right decision.

The conditional utilities used in expected-utility analysis have suppositions that direct evaluation to an appropriate world or set of worlds. The
theory of conditional utility introduces the interpretation of $U(o$ given $s)$ in
the expected-utility principle. In the conditional utility's expression, "given
s" is taken as short for "given s's truth." The condition's statement is
moodless. The condition supposes, roughly, that either s is the case or
would be the case if o were realized. Section 6.9 further explains this
interpretation of conditional utility and uses it in expected-utility analysis's
application to causal utility.

2.7 Justifying separation

The traditional expected-utility principle states a norm of rationality. This
chapter takes the principle to govern utilities interpreted as rational degrees
of desire and to advance an analysis of an option's utility. The principle
divides an option's utility according to possible outcomes taken as possible
worlds, or sets of possible worlds. This section uses principles of rationality
to justify for an ideal agent the expected-utility principle and the separation
of considerations it advances.

The expected-utility principle expresses utility's additivity. According to world-Bayesianism, the basic version of the principle, an option's utility is a sum of the utilities of the chances it creates for the possible worlds that might be the option's outcome. An equivalent version of the principle uses any partition of states and represents an option's outcome given a state with a set of possible worlds. An option offers a combination of chances for its exclusive possible outcomes. A proposition expressing possession of a chance for a possible outcome represents the chance, and a proposition expressing possession of a combination of chances similarly represents the combination of chances. Imagine betting a dollar that a coin toss turns up heads. The utility of a chance for a dollar gain given heads and the utility of a chance for a dollar loss given tails sum to the utility of the combination of chances the bet offers, that is, the bet's utility.

The expected-utility principle separates the utilities of the chances for outcomes that an option generates. The argument for separability of utilities of chances observes that the reasons for the utility of a chance that depends on p are independent of the reasons for the utility of a chance that depends on $\sim p$ because the propositions p and $\sim p$ are exclusive. The argument must, however, eliminate the possibility that changing a chance affects an option's utility through interaction with other chances.

Utilities are rational degrees of desire, and their additivity arises not from a convention of measurement but from principles of rationality. A normative argument for their additivity starts with utilities already introduced and shows that for certain combinations of chances, the utility of a combination equals the sum of its components' utilities. In science, the case for the combined gas law starts with the previously introduced quantities temperature (T), volume (V), and pressure (P) and uses empirical evidence to argue that $P = kT/V$ for a constant k. Normative utility theory starts with the previously introduced quantity degree of desire and uses principles of rationality to argue that rational degrees of desire obey the expected-utility principle. In neither case do the equations hold because of conventions of measurement. Utilities are a rational ideal agent's degrees of desire, and these degrees of desire order chances for exclusive possible outcomes and their combinations. The conventions of a utility-representation of the orders of chances and of their combinations do not make utilities additive. Utility's additivity follows from rationality's constraints on an ideal agent's degrees of desire.

The expected-utility principle's normativity disqualifies some methods of arguing for utility's additivity. Additive separability of possible outcomes grounds an additive utility-representation of preferences among options and possible outcomes according to which $U(o) = EU(o)$, as the expected-utility principle claims. This support for an additive representation does not constitute a justification of the expected-utility principle, however, because it simply assumes additive separability of possible outcomes, and because it shows only that the principle holds given some utility function's representation of preferences and not that it holds taking utilities as degrees of desire.

In utility theory, standard representation theorems, such as Savage's ([1954] 1972), show that satisfaction of certain axioms of preference ensures the existence and uniqueness of probability and utility functions that represent preferences as agreeing with expected utilities. The representation separates an option's utility into utilities of chances for exclusive possible outcomes, but only with respect to probability and utility functions constructed from preferences assuming utility's additivity. Hence, utility's additivity according to the constructed utility function is definitional, not normative.

Measurement theory uses representation theorems not to establish separability but to ground inference of quantities assuming separability. Commonly, a method of measurement assumes separability for a quantity, such as weight, by assuming that for the quantity some operation, such as placing together on a scale, applied to objects with the quantity yields a combination with the quantity such that the combination's amount of the quantity equals the sum of the objects' amounts of the quantity. Such an operation is called *concatenation*. Adopting an operation as concatenation for a quantity and representing it with addition does not justify the quantity's additivity unless the operation's adoption as concatenation has a justification. This section justifies combining chances for exclusive possible outcomes as a concatenation operation for utility and so justifies additivity and hence separability of the chances' utilities.

Section 2.5 presents several equivalent versions of the expected-utility principle. This section argues for a version of the principle that uses a partition of states to obtain an option's expected utility. The utility of a chance an option o offers if a state s holds is $P(s \text{ given } o)U(o \text{ given } s)$. A quantity equivalent to $U(o \text{ given } s)$ is $U(o\text{'s outcome given } s)$, where a set of possible worlds in which o and s hold or a disjunction of these possible worlds represents o's outcome given s. To support the expected-utility principle, this section argues for a binary principle that implies the expected-utility principle. As does the expected-utility principle, the binary principle treats chances for comprehensive outcomes. The binary principle asserts that the utility of a combination of two chances for exclusive outcomes (not necessarily exhaustive) equals the sum of the utilities of the chances. Two chances for exclusive outcomes that are not exhaustive combine to form a chance for the disjunction of the exclusive outcomes. The binary principle yields the expected-utility principle given that the chances for exclusive and exhaustive outcomes an option offers combine two into one until binary combination generates the whole combination of chances the option offers.

The binary principle expresses utility's additivity for a pair of chances for exclusive outcomes, and the expected-utility principle expresses utility's additivity for all the chances for exclusive and exhaustive outcomes that an option offers. When this section for brevity speaks of utility's additivity, it means additivity of utilities of chances for exclusive outcomes, either binary or exhaustive, unless it specifies another type of additivity.[17]

The argument for utility's additivity is not a proof but a collection of supporting points, as is typical for an argument justifying a basic normative principle. The first point in additivity's favor is the intuition, expressed by utility theorists such as Daniel Bernoulli (1738 [1954]), that the utility of a combination of two chances for exclusive outcomes equals the sum of the utilities of the chances. The sum is neither too high nor too low, but just right. Intuitions about cases corroborate. Granting that the utility of a chance for a dollar if heads turns up on a coin toss equals one-half the utility of gaining a dollar, and that the utility of a chance for a dollar if tails turns up on the same coin toss also equals one-half the utility of gaining a dollar, then the utility of the combination of chances is the sum of the utilities of the two chances. Because the utilities of the chances are equal, having both chances is twice as good as having just one. Their combination has the utility of gaining a dollar.

Standard philosophical methods, as Pust (2012) explains, use intuitions to support normative claims. The methods use reflective and informed judgments about what ought to be the case and not experimental findings about common human judgment and behavior. Psychological studies, such as those that Kahneman and Tversky (1979) report, typically target judgments of rationality for humans if they target judgments rather than behavior. Such studies show that subjects, given an opportunity, often

---

[17] Additivity's restriction to chances for exclusive outcomes ensures that the outcomes are not complementary. No two hold together, so none complements another. The outcomes' independence gives chances for them a type of independence. Additivity may generalize to certain types of independent chances for non-exclusive outcomes, namely, chances whose utilities are independent. Suppose that an agent is neutral toward risk and has a linear assignment of utility to money. Then, the utilities of two chances for a dollar gain from heads on independent coin tosses are additive. This section, however, does not treat additivity for chances for non-exclusive outcomes.

correct violations of normative principles of probability theory and decision theory. Initial judgments of rationality, when they violate the principles, generally are not reflective judgments. Subjects who persist in preferences that violate the sure-thing principle, formulated to handle only monetary consequences and the like, do not violate a revision of the principle that accommodates comprehensive consequences, as in Section 2.4, granting a plausible aversion to risk. In any case, for utility principles, such as the expected-utility principle, the relevant judgments concern rationality for ideal agents and not rationality for humans. Showing that intuitions support utility's additivity (as expressed by the expected-utility principle) makes a case, although not a decisive one, for utility's additivity, and this section supplements the case with other support for utility's additivity.[18]

The second point observes that in special cases, additivity follows from the formula for calculating a chance's utility. The formula states that the utility of a chance for an outcome is the product of the outcome's probability and its utility. This formula separates a chance's utility into two factors that it then combines by multiplication, a function with separable arguments. Suppose that an option offers chances for two exclusive outcomes of the same utility u. The combination of the two chances is the chance for the disjunction of their outcomes. The utility of the combination's outcome, the disjunction, is also u. Suppose that the probabilities of the two outcomes given the option are respectively p_1 and p_2. By probability's additivity, the probability of their disjunction, their combination's outcome, is $p_1 + p_2$. The formula for a chance's utility asserts that the two chances, respectively, have the utilities $p_1 u$ and $p_2 u$ and that their combination has the utility $(p_1 + p_2)u$. Because $(p_1 + p_2)u = p_1 u + p_2 u$, the utilities are additive. Imagine two chances for exclusive outcomes of unit utility. The probability of each outcome yields the utility of the chance for it. The additivity of the probabilities settles the additivity of the chances' utilities. The utility of the chances' combination equals the sum of their utilities. Imagine an option whose exclusive and exhaustive possible outcomes all have the same utility. Its utility similarly equals the sum of the utilities of the chances for its possible outcomes. Given the formula for the utility of chances, probability's additivity entails the additivity of the utilities of chances for equivalent outcomes.

Third, a traditional sure-loss argument supports utility's additivity. If an agent's utilities for chances fail to be additive, then the agent is subject

---

[18] Christensen (2004: 142), for example, endorses using intuition to support basic epistemic (normative) principles.

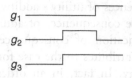

Figure 2.1 Gambles combined

to sure losses. Suppose that a gamble yields a dollar if an ace turns up on a roll of a die. Another gamble yields a dollar if a deuce turns up. A third gamble yields a dollar if an ace or a deuce turns up. If an agent is willing to pay more or less for the third gamble than the sum of the amounts he is willing to pay for the first gamble and for the second gamble, then he is open to a sure loss. The combination of the first two chances for a dollar is equivalent to the third chance for a dollar. Utility's additivity must hold to prevent vulnerability to sure losses from groups of bets.[19]

The sure-loss argument assumes the additivity of options' utilities. It assumes that the utility of two gambles equals the sum of the gambles' utilities and that a combination of gambles on exclusive propositions is equivalent to a gamble on the propositions' disjunction. Figure 2.1 represents gambles using the horizontal axis for outcomes and the vertical axis for their utilities. Because g_1 and g_2 pay off given exclusive outcomes, their combination g_3 pays off given the disjunction of these exclusive outcomes. Its utility equals the sum of the utilities of g_1 and g_2; that is, $U(g_3) = U(g_1) + U(g_2)$.

A simplified version of the sure-loss argument also assumes that the utility of money is linear so that the monetary value of two gambles equals the sum of their monetary values. The argument dramatizes the observation that a combination of gambles on exclusive propositions is equivalent to a gamble on the propositions' disjunction and so should have the same monetary value as the gamble on the disjunction, namely, the sum of the gambles' respective monetary values.

The sure-loss argument covers only cases in which its assumptions hold and is only as convincing as its assumptions. Because utility's additivity for options, its main assumption, is similar to utility's additivity for chances, the argument does not furnish conclusive support for utility's additivity. It is just part of the support for it.

---

[19] Ramsey (1931) and de Finetti (1937) construct sure-loss arguments supporting the axioms of probability. The style of argument extends to many normative principles, including the expected-utility principle.

Verification of consequences of utility's additivity constitutes the fourth point in its favor. Because consequences of additivity may hold without additivity's holding, verification of consequences does not conclusively establish additivity but contributes to the case for it. One consequence of additivity is commutativity. In fact, in an ordered combination of two chances for exclusive outcomes, the order of the chances does not affect the utility of their combination. The utility of a gamble that pays a dollar if rolling a die yields a five and a dollar if rolling the die yields a six, with chances taken in this order, equals the utility of a gamble that pays a dollar if rolling the die yields a six and a dollar if rolling the die yields a five, with chances taken in that order. The equivalence of the ordered combination of chances is so plain that evaluation typically does not distinguish the gambles according to the order of chances.

Another consequence of utility's additivity is its compositionality, which is equivalent to interchangeability of chances of equal utility. A substitution of an equivalent chance preserves the utility of a combination of chances as long as the substitution starts and finishes with a coherent combination of chances for exclusive outcomes. If the substitution moves from one coherent combination to another, then the combinations have the same utilities. A gamble that pays a dollar if tossing a coin yields heads has the same utility as a gamble that pays two dollars if drawing a playing card yields a diamond, because the utility of each chance for money equals one-half the utility of gaining a dollar, assuming that gaining two dollars has twice the utility of gaining one dollar. Also, given utility's indifference to the order of chances for exclusive outcomes, if two combinations have chances of identical utilities at each location under some ordering of chances, then by compositionality the utilities of the combinations are equal. In fact, a simultaneous set of exchanges of one equivalent chance for another that begins and ends with a coherent combination of chances preserves overall utility. For example, a gamble that pays a dollar if a coin toss yields heads becomes a gamble that pays a dollar if a coin toss yields tails after simultaneously exchanging the chance of a dollar if heads for a chance of a dollar if tails, and also the chance of nothing if tails for the chance of nothing if heads. The two gambles indeed have equal utilities, as compositionality requires.

According to utility's compositionality for combinations of chances, chances for exclusive outcomes are not complementary. However, complementarities between chances seem to arise in some cases. Take a chance for a dollar given heads on a coin toss and another chance for a dollar given tails on another coin toss. Substitute for the second chance an equivalent chance for a dollar if tails on the same coin toss that regulates the first chance. Then, the two chances yield certainty of a dollar, which, given aversion to chance, boosts utility. The two chances, one depending on heads and the other depending on tails, appear to be complementary when the same coin toss settles them. In fact, the case does not refute compositionality. Utility's compositionality for combinations of chances treats chances for finely individuated, comprehensive outcomes and not just for monetary outcomes, as Section 2.4 explains. Given an aversion to chance, a combination of chances generating exposure to chance has a comprehensive outcome that includes realization of the aversion to chance. Taking account of aversion to chance, the chances for money in the example change in relevant respects when implemented with one coin instead of with two coins. They involve nonequivalent outcomes. The outcomes of the implementation with two coins include realization of an aversion to chance, whereas the outcomes of the implementation with one coin do not. The implementation with one coin guarantees a dollar.

Measurement techniques move a chance from one context to another. However, a chance for a noncomprehensive outcome need not have the same utility in all contexts. Interchangeability requires chances for comprehensive outcomes that include, if relevant, nonmonetary factors such as realization of an aversion to risk. Moreover, if for an option some possible outcome includes a risk the option runs, then all the option's possible outcomes include the risk. A risk an option realizes is part of each of the option's possible outcomes. Consequently, not every replacement of a chance an option generates with a new chance yields an option. Suppose that the unreplaced chances an option generates are for outcomes that include a risk the option runs. If the new chance is for an outcome that does not include the same risk, then its combination with the unreplaced chances does not form an option. The chances for the possible outcomes together do not constitute an option's realization. Their combination is incoherent. Interchangeability governs only substitutions that go from one coherent combination to another. It holds provided that chances are for comprehensive outcomes and combinations are coherent.

Utility's additivity also entails utility's separability, in fact, its additive separability. Verification of this consequence of additivity begins with an explication of utility's separability. To start, note a difference between utility's additivity and its separability. Its additivity governs single vectors of utilities. In contrast, its separability governs sets of vectors of utilities. Separability of utilities of chances applies to the utilities of chances at locations in vectors of utilities of chances. Variables represent the locations.

Addition of probabilities and additive separability of probability variables create a model for addition of utilities of chances and additive separability of utility variables. The addition law for probability treats a disjunction of incompatible propositions. The probability of each disjunct contributes independently to the probability of their disjunction. Because of the additivity of probabilities, variables having probabilities of disjuncts as values, that is, locations for the probabilities in an ordered pair, are separable given constant evidence. The probability of a disjunction increases after replacing a disjunct whose probability occupies a location with another disjunct of greater probability. Although increasing the probability of one disjunct in $p \vee \sim p$ draws probability from the other, if a pair of incompatible disjuncts is not exhaustive, increasing the probability at the location for a disjunct by changing the disjunct in a way that preserves the incompatibility of the disjuncts does not draw probability from the other disjunct. The locations for probabilities of the disjuncts are mutually and so additively separable; an additive representation of the disjunctions' order is possible.

Because an option offers chances for exclusive and exhaustive outcomes, increasing the probability of an outcome decreases the probability of other outcomes. However, increasing a chance's utility by increasing the outcome's utility need not affect other chances' utilities. Moreover, in a pair of chances for exclusive but not exhaustive outcomes, increasing a chance's utility by increasing the outcome's probability in a way that preserves exclusivity does not affect the other chance's utility. Utility's separability in combinations of chances treats utilities at locations in vectors representing coherent combinations of chances.

Utility's additivity governs chances an option creates, and utility's additive separability governs variables whose values form the chances. Suppose that options o_1 and o_2 and exclusive and exhaustive states s_1 and s_2 generate the utilities of chances that Table 2.1 displays. The utility of a chance in a cell for a state and an option equals the product of the state's probability given the option and the option's utility given the state. If the states are worlds, then the utility of an option given a world equals the world's utility. According to utility's additivity, the utility of o_1 is the sum of the two utilities in the row for o_1. Similarly, the utility of o_2 is the sum of the two utilities in the row for o_2.

Table 2.1 *Utilities of chances*

	s_1	s_2
o_1	1	2
o_2	3	2

In the table, a column represents a variable whose value is a chance's utility. Filling the two columns with the values 1 and 2 represents option o_1. Filling the two columns with the values 3 and 2 represents option o_2. The left column is separable from the right column because the order of its values agrees with the order of the options after fixing the value of the right column. The table fixes it at 2. Utility's additivity implies that the left column is separable from the right column. Suppose that the table enters rows for more options. Utility's additivity implies that the left column and the right column are separable from each other and so additively separable. In general, in a decision problem, given a partition of states, utility's additivity for each option implies the additive separability of variables whose values are the utilities of chances that options generate with respect to the partition of states. The additivity of the utility of a combination of chances that an option offers makes, for variables yielding the chances' utilities, every set separable from its complement so that the variables are additively separable; an additive representation of the order of combinations is possible.[20]

Consider variables having the utilities of chances as values. Given utility's additivity, the utility of a chance makes the same contribution to the utility of any option that generates it. Take a 50 percent chance of a 10-unit gain in utility. It makes the same contribution to options' utilities for all options that yield it. It adds the same amount of utility to any combination to which it belongs. Hence, its utility is separable from the utilities of other chances. If the utility of losing a gamble is fixed, then increasing the utility of winning the gamble by, say, increasing the prize, increases the gamble's utility.[21]

---

[20] This separability permits simplifying choices by ignoring states in which the utilities of chances that options generate are the same. The occasion for this simplification arises rarely. In most decision problems and for most partitions of states, in each state the options produce chances with different utilities.

[21] Suppose that one chance in a combination may influence another chance's utility without changing the chance. In this case, each combination's utility may equal the sum of its component's utilities although the order of combinations differs from the order of a component variable's values. An increase in its value may lower the utilities of values of other component variables and so lower the utility of the new combination. Imagine that the vectors of chances (c_1, c_2) and (c_3, c_2) represent combinations of chances, and the corresponding utility vectors are $(1, 2)$ and $(2, 0)$. Changing the first location's value in the first vector of chances from c_1 to c_3 increases the first location's value in the corresponding vector of utilities from 1 to 2. However, the change in the vector of chances lowers the second location's value c_2 from 2 to 0, as the corresponding vector of utilities shows. According to additivity, the order of vectors of chances goes by the sum of the corresponding vectors of utilities, so the first vector of chances ranks higher than the second vector of chances. Therefore, the first location of the vector of chances is not separable from the second location. The chance c_3 ranks higher than the chance c_1, but (c_1, c_2) ranks higher than (c_3, c_2). The order of values of the first location in the chance vectors does not agree with the order of vectors, holding fixed the value of the second location. This

Intuition confirms utility's additive separability. Changing a chance in a combination of chances to increase the utility at a location in a vector of utilities of chances in the combination increases the combination's utility, assuming that the combination remains coherent after the change. For example, improving a gamble for a dollar gain if a roll of a die yields a six by having the gamble pay a dollar if the roll yields either a five or a six increases the gamble's utility. Also, improving the gamble by increasing the possible gain from a dollar to two dollars increases the gamble's utility.

This section's four points together make a strong case for utility's additivity for chances. They endorse the expected-utility principle. Because at some points the argument for utility's additivity for chances uses probability's additivity and utility's additivity for options, the argument does not support utility's additivity for chances as fundamental. Hence, the separability that follows from utility's additivity for chances is not fundamental. In contrast, Chapter 3 argues for a fundamental type of additivity and separability for utility.

This chapter argued for a traditional version of the expect-utility principle that calculates an option's expected utility using probability taken as rational degree of belief and utility taken as rational degree of desire. The traditional principle is normative, in contrast with an alternative version that uses probability and utility functions defined to represent preferences "as if" following expected utilities so that an option's utility equals its expected utility by definition. The traditional principle need not restrict outcomes, so they may occur in many contexts, to facilitate constructing probability and utility functions from preferences. For accuracy's sake, the chapter takes an option's outcome comprehensively so that it includes everything that matters and, for generality's sake, formulates the expected-utility principle so that it applies to the various types of utility that subsequent chapters introduce.

---

counterexample to additivity's entailment of separability of chances does not arise if the change in c_2's utility changes c_2 to another chance. Additivity entails separability of chances given that changing the probability or utility of a chance's outcome changes the chance. Then any chance's influence on another chance's utility changes the other chance. This section individuates chances according to outcomes and their probabilities rather than according to outcomes and their probabilities and utilities. Given its coarse individuation of chances, utility's additivity entails separability of utilities of chances but not also separability of chances.

CHAPTER 3

Intrinsic utility

A division of a world into parts and a method of evaluating the parts ground
a world's evaluation. This chapter presents the elements of a world's utility
and establishes their separability as a norm for ideal agents. The elements'
independence supports not only their separability but also their additivity.
It justifies a method of analyzing a world's utility.

The method of utility analysis uses basic intrinsic attitudes. After explaining these attitudes, the chapter argues that their independence supports a
norm of independence for evaluations of their realizations. The independence of these evaluations supports separability and additivity for a type of
utility, intrinsic utility. Intrinsic utility's additivity grounds the chapter's
method of utility analysis, which in turn grounds the following chapters'
methods of simplifying choices.

3.1   Pros and cons

A job seeker prizes more a satisfying job with security than a satisfying job
without security. Realization of two goals instead of just one makes the first
job more appealing. The job seeker combines evaluations of satisfying work
and of security to evaluate an outcome with both. Deliberations may derive
a world's utility from evaluations of its parts.

A world has certain separable features that form the atoms of its utility's
analysis. This chapter defines those features using the distinction between
intrinsic desires such as the desire for pleasure and extrinsic desires such as
the desire for wealth, and a corresponding distinction between intrinsic and
extrinsic aversions. Then, it advances a version of traditional evaluation by
adding pros and cons: *a world's utility is a sum of quantitative evaluations of
its realizations of basic intrinsic desires and aversions.*

An option's comprehensive upshot is a possible world. A world's utility
stems from the pros and cons of the world's realization. If the world's
realization satisfies a desire, that is a pro. If it realizes an aversion's object,

that is a con. A world's utility comes from tallying its realization's pros and cons. The addition must weight each pro and con according to its strength. Realization of a strong desire has more weight than realization of a weak desire. The addition must also neither omit nor double count considerations. It should count each relevant consideration exactly once. If a world's realization entails an aversion's realization, its utility must consider the aversion's realization. If a world's realization entails realization of a desire to work for NASA and also realization of a desire to work for the National Aeronautics and Space Administration, it may not count the realizations of both desires – their objects are the same although the desires represent them differently. Using basic intrinsic desires and aversions covers all pros and cons without double counting.

Holism maintains that a world's utility does not separate into the utilities of the world's segments. However, it has no proof, for all types of a world's utility, that all types of separation fail. The literature presents only counter-examples to some types of separation for some types of utility, such as counterexamples that discredit separation of a world's comprehensive utility into its parts' comprehensive utilities. This chapter introduces a type of utility for which separability holds. The type of utility has narrower scope than has comprehensive utility.

People analyze an option's utility according to the goals it realizes, although classical decision theory does not incorporate such analyses. This chapter extends classical decision theory by showing how to evaluate an option's outcome using the goals it realizes. To ground efficient forms of utility analysis, the chapter extends the method of evaluation to propositions representing proper parts of worlds. The method of evaluation yields a type of utility with narrow scope that the chapter calls *intrinsic utility*. The elements of intrinsic utility are realizations of basic intrinsic desires and aversions.

3.2   Intrinsic desire and aversion

This section explains the intrinsic desires and aversions that are the source of intrinsic utilities. These mental states form the basis of all an agent's desires and aversions. Their treatment in philosophy goes back to Plato's *Republic*.

Section 2.2's introduction of utility includes a characterization of desire. The opposite of desire is aversion, and between desire and aversion lies indifference. Desire, aversion, and indifference are propositional attitudes and so have propositions as objects. For simplicity, colloquial expressions of a desire use an infinitive phrase or noun phrase instead of a that-clause stating the proposition the desire targets. For example, Mary may say John desires to win instead of saying he desires that he win. Also, although a desire has as its object a proposition's realization, one often says its object is the proposition itself. Similar conventions apply to expressions of aversions and attitudes of indifference.

A possible world is a maximal consistent specification of matters relevant to desires and aversions. It is maximal in the sense of specifying for every goal whether that goal is realized. Sets of possible worlds represent propositions well in many contexts, although the representation conflates some distinctions between propositions, for example, the distinction between the proposition that p and the logically equivalent proposition that $(p \text{ \& } q) \text{ } v$ $(p \text{ \& } \sim q)$. The conflation of logically equivalent propositions does not harm this book's project because the project targets a rational ideal agent who gives logically equivalent propositions the same evaluation. A conflation of necessary truths that are not a priori equivalent causes trouble, however. Suppose that God exists necessarily. Then, the set of all possible worlds represents the proposition that God exists and also the proposition that God exists or does not exist. A rational ideal agent, given the rationality of agnosticism, may assign different utilities to the two propositions. In such special cases, the book preserves the difference between the propositions despite the identity of the sets of worlds that represent them.

Propositions are fine-grained objects of desire. Oedipus wants to marry Jocasta but not his mother, although unknown to him Jocasta is his mother. According to some accounts of events, Oedipus's marrying Jocasta is the same event as his marrying his mother. Propositions do not represent events in that sense. The proposition that Oedipus marries Jocasta differs from the proposition that he marries his mother. Oedipus may desire that the first, but not that the second, proposition be true. In fact, a desire that a proposition hold obtains relative to a way of understanding the proposition. A traveler may want to go to St. Petersburg and not to Leningrad, although the proposition that he goes to St. Petersburg is the same as the proposition that he goes to Leningrad. This chapter, for simplicity, assumes that propositions are objects of desire. It keeps on the shelf the possibility of distinguishing utilities of the same proposition given two ways of understanding the proposition.

Utility's dependence on propositions' expressions creates a complication rather than a problem for principles of utility. A proposition's utility may be relativized to a way of understanding the proposition, and utility principles may treat utilities so relativized. To highlight utility analysis's main features, this book treats only cases that do not require relativization. Both a realist interpretation of utility and a constructivist interpretation of utility generate this complication; as Kahneman and Tversky's (1979) study of framing effects shows, a person's preferences between options depend on the person's way of understanding the options. Because ignorance of empirical facts may make preferences sensitive to propositions' expressions, even utility principles for ideal agents face the complication.

Some things we desire for their own sakes and other things as means to things we desire for their own sakes. According to tradition, desires of the first type are called intrinsic, and desires of the second type extrinsic. In typical cases, pleasure is an object of intrinsic desire, and money is an object of extrinsic desire. A similar distinction applies to aversions.[1]

A desire is intrinsic just in case it obtains because of its object's logical consequences (in the broad sense of its a priori consequences or implications). An intrinsic desire that a proposition hold considers only the proposition's logical consequences. In contrast, an extrinsic desire that a proposition hold considers all that follows from the proposition's truth, including its causal consequences. Intrinsic aversion and indifference have analogous characterizations. For simplicity, this chapter ignores intrinsic indifference (except to set a zero point for intrinsic utility) because realizations of its objects do not directly affect a world's attractiveness.

An intrinsic attitude evaluates intrinsic properties of its object's realization and is not intrinsic in virtue of the attitude's intrinsic properties. An intrinsic desire is not a desire that is intrinsic, but rather a desire whose grounds are intrinsic features of its object's realization. Its classification depends on the scope of its evaluation of a proposition's realization. Hence, an agent may have both an intrinsic and an extrinsic desire concerning the same proposition. For example, an agent may desire to be healthy for the sake of being healthy and also because being healthy is a means of being productive. Intrinsic desire and aversion evaluate the intrinsic features of a proposition's realization taken as the proposition's logical consequences. Thus, a person's intrinsic desire to be healthy and wise attends only to the logical consequences of being healthy and wise (such as being wise) and not also to wisdom's monetary rewards. Both intrinsic and extrinsic conative

---

[1] An agent may desire a college education for his own sake, for his mother's sake, for the sake of scholarship, and for the sake of improving the world. These examples of desiring for the sake of something do not illustrate the way the introduction of intrinsic desire uses desiring something for its own sake. A proposition desired to be true does not have an interest to be promoted, nor does its realization. One does not desire the proposition's realization to promote the proposition's interest or its realization's interest when one desires it for its own sake. The usual characterization of intrinsic desire uses desiring for its own sake to indicate an end, and not to indicate an interest the desire serves. Weirich (2001: chap. 2; 2009) and Schroeder (2004: 5, 132) offer accounts of intrinsic desires and aversions.

attitudes attend to everything in their scope although they have different scopes. Consequently, an extrinsic desire to exercise takes account of all logical and causal consequences of exercise, whereas an intrinsic desire to be wise takes account of all and only logical consequences of being wise.

Because intrinsic desire involves subjective evaluation, a fully rational agent may at one time have an intrinsic desire to bask in the sun and at another time lack that intrinsic desire, even though the logical consequences of basking in the sun are constant. However, intrinsic desires are generally more constant than are extrinsic desires because information about causal consequences and circumstances of evaluation do not provide reasons for them.

Grant that every desire holds for the sake of something. It can hold for the sake of the desire's object or for something else; the two possibilities are exhaustive. If the former, the desire is intrinsic; if the latter, it is extrinsic. So intrinsic and extrinsic desires exhaust desires. Perhaps the desire to keep one's wedding ring seems neither intrinsic nor extrinsic. However, in a typical case, possessing the ring gives one a symbol of one's marriage, and one intrinsically desires having such a symbol. So the desire to keep it is extrinsic.

Sober and Wilson (1998: 217–22) distinguish ultimate and instrumental desires. They note that an instrumental desire for the means to an end ceases with satisfaction of the ultimate desire for the end, but the ultimate desire for an end outlasts satisfaction of an instrumental desire for a means to the end until the means achieves the end. Satisfying intrinsic desires similarly extinguishes extrinsic desires for a means of satisfying them. Ultimate and instrumental desires correspond to intrinsic and extrinsic desires, respectively.

In psychology, an agent's desire counts as intrinsic if it is stable, long-standing, and largely beyond cognitive influence. In philosophy, it counts as intrinsic if the agent would want its realization even if that brought nothing else that the agent wants. These are rough characterizations that a theory of intrinsic desire refines.

Are the objects of intrinsic desires the same as final ends, that is, non-derivative ends? Final ends are not pursued for the sake of other ends. So, they are objects of intrinsic desire. However, objects of intrinsic desires need not be final ends. One may desire health and wisdom intrinsically, although the desire for the combination derives from the desire for each component and so is not a final end. Intrinsic and final ends are not the same because one intrinsic end may derive from another.[2]

---

[2] Korsgaard (1983) warns against conflating the distinctions between an intrinsic good and an extrinsic good, between a final good and a derivative good, between an unconditional good and a conditional good, and between a noncontingent good and a contingent good. Unconditional goods are good

Intrinsic desire is not part of the stock in trade of decision theory. Decision theory neglects it because of a proclivity for operationalism. However, as the book's Introduction notes, operationalist theories of meaning are implausible. A plausible operational standard for theoretical entities insists only on evidence of theoretical entities. Intrinsic desires meet this inferential standard. In a rational ideal agent, a desire's independence of information indicates its being an intrinsic desire. This rough test of intrinsic desire licenses its inclusion in decision theory.[3]

Because an intrinsic desire or aversion may derive from other intrinsic desires or aversions, an evaluation of a world using realizations of intrinsic desires and aversions risks double counting considerations. Suppose that an intrinsic desire for pleasure and wisdom rests on an intrinsic desire for pleasure and on an intrinsic desire for wisdom. An evaluation of a world with both pleasure and wisdom double counts pleasure and double counts wisdom if it sums evaluations of pleasure, wisdom, and pleasure and wisdom. To solve the problem, a world's evaluation attends only to realizations of basic intrinsic desires and basic intrinsic aversions. An agent's intrinsic desire is basic only if no intrinsic desire is among the agent's reasons for it. A basic intrinsic aversion meets an analogous condition. A person typically has a basic intrinsic desire for pleasure and a basic intrinsic aversion to pain. Given an intrinsic desire for pleasure and an intrinsic desire for wisdom, an intrinsic desire for pleasure and wisdom is not basic. The intrinsic desire for pleasure and the intrinsic desire for wisdom furnish reasons for intrinsically desiring the combination of pleasure and wisdom.

Intrinsic desires and aversions are intrinsic (conative) attitudes. Intrinsic attitudes come with intensities. The reasons for an intrinsic attitude include the reasons for its intensity, in fact, the reasons for all its conative features, not just the reasons for its existence. A basic intrinsic attitude, BIT, may furnish a reason for the intensity of a nonbasic intrinsic attitude as well as a reason for having an intrinsic attitude toward the attitude's object.[4] An agent's intrinsic attitude is *basic* if and only if no intrinsic attitude is among

---

independently of conditions, whereas conditional goods are not. Noncontingent goods are good independently of contingencies, whereas contingent goods are not. Some intrinsic desires are unconditional (with respect to information and circumstances), noncontingent (with respect to possibilities), and final (with respect to desires). An intrinsic good is good because of its intrinsic features; its unconditionality and noncontingency derive from its being an intrinsic good.

[3] People have an intrinsic desire for happiness. Positive psychology studies happiness, which it calls subjective well-being. Kahneman et al. (2003) treat especially the measurement of subjective well-being.

[4] "BIT" instead of "BIA" is the abbreviation for "basic intrinsic attitude." Using the second letter of "attitude" makes the abbreviation a single syllable and makes it suggest the mental state's role in evaluations.

the agent's reasons for holding the attitude. This definition of basicness is theoretically fruitful because according to it, basic intrinsic attitudes ground principles of utility analysis that simplify choices.

An *effective reason* is an agent's reason for doing something or being in some state. It is an internal rather than an external reason. A BIT is an intrinsic attitude for which no intrinsic attitude furnishes an effective reason, that is, a reason that motivates the agent to hold the intrinsic attitude.

In most cases, new information does not affect an agent's basic intrinsic attitudes. However, in some cases, it may change them because they depend on the agent's conception of their objects. Information may (as a pill may) change an agent's attitude to a proposition without introducing reasons to change the attitude. In particular, because it enables new thoughts, new information may trigger a change in BITs. Tasting pineapple for the first time provides information that may prompt a basic intrinsic desire that one taste pineapple. The new information does not provide a reason for the BIT, however. If one has a taste for sweetness and discovers that pineapples are sweet and so forms a desire to taste pineapple, then the new desire is derived rather than basic. Because a BIT is held independently of other intrinsic attitudes, new information about the chances of satisfying other intrinsic attitudes does not furnish reasons for changing the BIT. A BIT's independence gives it stability.

Parfit (1984: 165–70) observes that an intrinsic aversion to a pain lying in the future is typically greater than an intrinsic aversion to a similar pain already experienced. The temporal relation to the experience affects the aversion. Although a full theory of basic intrinsic attitudes investigates their origin, evolution, and extinction, this chapter does not inventory factors affecting BITs because it just uses an agent's current BITs to ground an analysis of the agent's current utility assignments.

An ideal agent may not have an intrinsic attitude toward every proposition. The agent may skip inconsistent propositions. Also, a disjunction of the object of a basic intrinsic desire and the object of a basic intrinsic aversion may not be the object of an intrinsic attitude. An agent may suspend intrinsic evaluation of the disjunctive proposition.

As Schroeder (2004: 132–33) observes, desire that p and aversion toward not-p are psychologically distinct. Rationality imposes consistency constraints for the two attitudes, such as not simultaneously desiring that p all things considered and having an aversion toward p all things considered. Also, in typical cases, given a desire that p, rationality requires an aversion toward not-p. A rich normative theory distinguishes psychological states to formulate norms connecting the states. It includes requirements of rationality concerning the relation between desire and aversion rather than defining aversion in terms of desire so that their relation is not normative.

The negation of a BIT's object need not be the object of a BIT. A basic intrinsic desire that p need not generate a basic intrinsic aversion toward $\sim p$. An irrational agent clearly may have the basic intrinsic desire without the basic intrinsic aversion. Also, a rational agent may have the desire without the aversion. A rational agent may have a basic intrinsic desire for the pleasure of chocolate ice cream without a basic intrinsic aversion to the pleasure's absence; the pleasure's absence does not imply the absence of all pleasure. Moreover, if an agent's reason for an intrinsic aversion toward $\sim p$ is an intrinsic desire that p, then the agent has a derived rather than a basic intrinsic aversion toward $\sim p$. The nonrealization of a BIT is generally not the object of a BIT because the reasons concerning an agent's attitude to the nonrealization make the attitude derivative rather than basic.

One BIT's realization may entail another BIT's nonrealization. For example, an agent may have basic intrinsic desires for pleasures of various degrees. Having now pleasure of a certain degree entails not having now pleasure of a different degree. So, for such an agent, one BIT's realization entails another's nonrealization.

Can one BIT's realization entail another BIT's realization? Having pleasure for two minutes entails having it for one minute. An agent may have intrinsic desires for both states, and if both intrinsic desires are BITs, then realizing the first BIT entails realizing the second BIT. However, an intrinsic desire for one minute of pleasure supplies a reason for an intrinsic desire for two minutes of pleasure. Hence, in a rational ideal agent, the latter is not a BIT. In general, for a rational ideal agent, no BIT's realization entails another BIT's realization. If one intrinsic attitude's realization entails another intrinsic attitude's realization, then the latter is a reason for the former so that the former is not a BIT.

An agent may desire pleasure during an interval, however short, as a means of experiencing pleasure during a longer interval. However, for a typical agent, an intrinsic desire for pleasure during an interval ceases if the interval becomes very short. For a rational ideal agent with an intrinsic desire for pleasure during some period of time, I assume that some shortest period of time exists such that the agent intrinsically desires pleasure during it. That intrinsic desire constitutes a BIT, and it furnishes a reason for pleasure during longer periods. The shortest period may be the duration of a just noticeable experience of pleasure. No shortening of the experience yields a conscious experience of pleasure. For a rational ideal agent, able to experience pleasure consciously during periods smaller and smaller, with no limit other than zero duration, BITs grounding intrinsic desires for pleasure may not exist. I put aside such agents.

Entailments among BITs' objects, if they were to hold, would block additivity of the objects' intrinsic utilities. If a conjunction and a conjunct were both objects of BITs, then the conjunction's intrinsic utility would typically differ from the sum of the intrinsic utilities of the BITs it realizes. Summing the intrinsic utilities of the conjunction and the conjunct double counts the conjunct and so inflates the sum. Because entailments do not arise, intrinsic utility's additivity does not need a restriction to cases in which no BIT's object entails another BIT's object.

As noted, an agent's reasons for holding an intrinsic attitude are not objective, external reasons of which she may be unaware, but subjective, internal reasons accessible to her. An agent's reasons for holding an intrinsic attitude are reasons that cause the attitude. An agent's basic intrinsic desires and aversions furnish reasons for other intrinsic attitudes and, in a rational ideal agent, cause the agent's other intrinsic desires and aversions in virtue of the agent's grasp of the attitudes' objects. Reasons are causes in rational ideal agents because such agents ignore no reasons they possess. For example, if the intrinsic desires to be healthy and to be wise are basic, then they cause a nonbasic intrinsic desire to be healthy and wise. A rational ideal agent's basic intrinsic attitudes are a causal foundation for her intrinsic attitudes. Because an ideal agent's reasons are causes, the agent's reasons for intrinsic attitudes do not form a circle. An ideal agent's reasons for holding intrinsic attitudes have a tree-like structure.[5]

May intrinsic desires that p and that $(p \& q)$ be mutually supportive? An agent may intrinsically desire that $(p \& q)$ because of an intrinsic desire that p. However, an intrinsic desire that p holds because of the implications of p. The agent's intrinsic desire that $(p \& q)$ does not support an intrinsic desire that p. It supports only an extrinsic desire that p as necessary for realization of $(p \& q)$ (the proposition that p may be the object of both an intrinsic desire and an extrinsic desire). Suppose that an agent desires intrinsically that his sister prosper and desires intrinsically that his sister and her husband prosper. His desire that the couple prosper is not a reason to desire intrinsically that his sister prosper. It is just a reason to desire

---

[5] Mele (2003: chap. 2) defends the view that an agent's reasons for an act are causes of the act. Counterfactual dependency is a sign that the reasons are causes of the act. In a typical case, if the reasons had been absent, then the act would not have occurred. Similar evidence suggests that an agent's reasons for an attitude are causes of the attitude.

extrinsically that she prosper as a means of the couple's prospering. Although the two desires are mutually supportive, the two intrinsic desires are not mutually supportive.

Not every reason for an intrinsic attitude is an intrinsic attitude. A reason for a BIT may be an experience. Even in a rational ideal agent, not every cause of an intrinsic desire is a reason for the intrinsic desire. Almost anything, such as a blow to the head, may cause a desire: (1) A person's extrinsic desire for money may prompt thoughts about having the wisdom to spend well. These thoughts about wisdom may then produce an intrinsic desire for wisdom. (2) An agent's intrinsic desire for health may lead to thoughts about having the wisdom to live well, which then produce an intrinsic desire for wisdom. (3) An agent's intrinsic desire for beauty may excite him and so incidentally intensify his intrinsic desire for wisdom. These cases are not counterexamples to the section's account of BITs.[6]

Intrinsic attitudes that are an agent's effective reasons (good, bad, or indifferent) for other intrinsic attitudes meet three conditions. First, they are sufficient causes of those attitudes given the agent's background psychological condition with respect to factors besides intrinsic attitudes and fully account for the intensity of those attitudes. Second, no other types of attitudes intervene, so they are proximate causes. Third, their ceasing now would immediately undermine the attitudes they cause, so they are sustaining causes. An agent's BITs are his effective reasons for his other intrinsic attitudes, and no intrinsic attitudes are his effective reasons for any of his BITs. Consequently, BITs are sufficient, proximate, sustaining causes of other intrinsic attitudes, and no intrinsic attitudes are sufficient, proximate, sustaining causes of any BIT.[7]

Given the account of BITs, causally anomalous cases are not counterexamples. They are not cases in which the effective reasons for a BIT are some other intrinsic attitudes. They feature causes for a BIT that are not intrinsic attitudes, or that are not sufficient, proximate, sustaining causes, or that in some other way fail to be effective reasons for the BIT. For instance, a desire for money may cause a BIT for wisdom but in a rational ideal agent

---

[6] I thank Wlodek Rabinowicz (personal communication, 2002) and James Joyce (2003) for good questions about the causes of BITs.

[7] The relation between basic intrinsic attitudes and other intrinsic attitudes is a basing relation for intrinsic attitudes. The basing relation among intrinsic attitudes involves the agent's reasons for holding the intrinsic attitudes. This section does not analyze the basing relation among intrinsic attitudes. The basing relation for intrinsic attitudes is similar to epistemology's basing relation for beliefs. The basing relation for beliefs concerns reasons for which a belief is held. It often receives a causal analysis. Korcz (2002) characterizes the relation and reviews the literature on it. McCain (2012) offers a recent account.

is not a sufficient, proximate, sustaining cause of that BIT. The BIT for wisdom survives the desire for money's waning. The desire for money is not an effective reason for the BIT.

Harman (1967: 804) objects to the assumption of a single set of intrinsic desires grounding all intrinsic desires. The choice of a foundational set is arbitrary, he claims. May not various sets of intrinsic attitudes yield the whole set of intrinsic attitudes, as various sets of axioms may yield the whole set of logical truths in propositional logic? Defining BITs using an agent's reasons for intrinsic attitudes rather than using logical relations among the attitudes' objects removes the worry. Although the contents of intrinsic attitudes may allow for taking different sets as basic, an agent's reasons for having intrinsic attitudes make a single set basic. Multiple systems of BITs grounding an agent's intrinsic attitudes may be compatible with available evidence of the agent's preferences among acts. However, that is just a problem of measurement and not a problem concerning the structure of intrinsic attitudes and the existence of BITs. BITs may exist although data about the agent's preferences are insufficient for identifying the BITs.

Rational ideal agents respond to their reasons. Imagine an agent with multiple sufficient reasons for an intrinsic attitude. Her intrinsic attitude is overdetermined (unless her efficiency in reasoning prevents overdetermination). Nonetheless, a unique set of BITs furnishes her reasons for holding her intrinsic attitude. Her reasons for holding the attitude are the reasons that produce it. Consider an intrinsic desire that x & y & z. Suppose that intrinsic desires that x & y and that z justify it, and also intrinsic desires that x and that y & z. Despite the overdetermination of reasons, the basic intrinsic desires grounding the intrinsic desire that x & y & z are unique. If the intrinsic desires that x, that y, and that z are basic, then they ground the intrinsic desire that x & y & z.

Suppose that the foregoing argument is wrong, and the set of BITs is arbitrary. Choice of the set of BITs for the separation of a world's intrinsic utility is then conventional, as is the choice of the unit for intrinsic utility. The arbitrariness does not matter as long as each complete set of BITs generates the same intrinsic utilities for worlds. Then, any set of BITs grounds a separation of a world's intrinsic utility. The separation does not require a unique set of BITs.

This chapter makes some assumptions about BITs. An intrinsic desire for luxury may cause an irrational desire for a luxury car and may prompt irrational overspending on a car. BITs may cause irrational desires and irrational acts. The chapter's idealizations put aside such cases; the chapter assumes a fully rational ideal agent. In such an agent, BITs are rational and do not cause any irrational desires. Although an account of rational BITs is necessary for a full account of rational choice, this chapter just assumes that in the cases it treats, BITs are rational in virtue of compliance with all pertinent principles of rationality.

Not every intrinsic desire and aversion is quantitative. For example, compare the proposition that one experiences pleasure with the proposition that one experiences now the pleasure of a sip of water. The first proposition is more vague than is the second. Perhaps, although one intrinsically desires the realizations of both propositions, only the desire for the second's realization is definite enough to be quantitative. To simplify, this chapter treats only cases in which BITs are quantitative. Their being quantitative entails their objects' comparability.

If basic intrinsic attitudes were to generate all other intrinsic attitudes but not all features of other intrinsic attitudes, then basic intrinsic attitudes would incompletely ground intrinsic attitudes. In that case, a basic intrinsic desire that p and a basic intrinsic desire that q might generate an intrinsic desire that p and q without the first two desires' intensities settling the third desire's intensity. Intrinsic-utility analysis assumes that basic intrinsic attitudes completely ground intrinsic attitudes so that realizations of BITs settle all intrinsic utilities. The chapter treats rational ideal agents with precise intrinsic attitudes that meet the assumption.[8]

An agent may have an infinite number of BITs, say a BIT for each intensity of pleasure at each moment of time. Just noticeable differences may reduce the intensities and times to a finite number, but the agent may care about the intensities and not just perceptions of them. This chapter's analysis of a world's utility puts aside such cases and assumes that an agent has a finite number of BITs.

Although utility analysis treats ideal agents, it ascribes to them attitudes that real people have. To explain the basic intrinsic attitudes of ideal agents, it discusses the basic intrinsic attitudes of real people. Also, utility analysis for ideal agents lays the groundwork for a general theory of utility that treats nonideal agents too. How may utility analysis generalize to treat humans? What are the basic intrinsic attitudes of real people?

---

[8] If a diner has a basic intrinsic desire for peaches and an intrinsic desire for peaches and cream, then because the first desire is a reason for the second, the second is not basic. However, if the diner has an intrinsic desire neither for cream nor for cream's complement to peaches, then perhaps basic intrinsic desires do not fully account for her intrinsic desire for peaches and cream. Suppose that the definition of a BIT were modified so that instead of being an intrinsic attitude not grounded in any intrinsic attitude, it were an intrinsic attitude not fully grounded in intrinsic attitudes. Then, BITs would multiply. Although they would fully account for intrinsic attitudes in all cases, their independence would be in jeopardy.

Given human cognitive limits, a large number of world parts that a person cares about generate by composition too many worlds to evaluate unless their evaluations depend on evaluations of their parts. Because people evaluate worlds despite caring about a large number of world parts, it is likely that their intrinsic attitudes toward world parts settle their intrinsic attitudes toward worlds. It is unlikely that their intrinsic attitudes are holistic and that their BITs attach to worlds. However, this section's account of BITs leaves open their objects' specificity. An agent's intrinsic attitude toward a possible world may not follow from her intrinsic attitudes toward the world's features so that her intrinsic attitude toward the world is basic. If an agent's intrinsic attitudes are holistic, then her BITs have worlds as objects. This book treats agents for whom the objects of BITs are not worlds but proper parts of worlds. It derives the consequences of having BITs attached to propositions less specific than worlds. The book does not advance normative principles requiring BITs of this type. So the norms of separability it presents rest on the assumption that the agents it treats have such BITs.

Applying this section's normative model of desires to a person requires psychological investigation to discover the person's BITs. How may a psychologist discover a person's BITs? Does a person have a basic intrinsic desire for health, or does that intrinsic desire derive from a basic intrinsic desire for health today? This section's model puts aside such psychological questions by leaving open the objects of BITs. The model treats an agent with BITs that are foundational for all the agent's intrinsic attitudes. It assumes the possibility of this structure of desires but does not make the psychological assumption that a human agent's desires have this structure. It postpones considering not only how the model applies to humans but also whether the model applies to humans. Future psychological research may show how to extend and apply the model to humans.[9]

3.3 Intrinsic utility's definition

A vacationer may intrinsically desire two hours at the beach twice as much as one hour. The relations of intrinsic attitudes may make them apt for quantitative representation. Intrinsic utility is a quantitative representation of intrinsic attitudes.

---

[9] Huang and Bargh (2014) study motivational structures involving goals that resemble basic intrinsic desires.

Because ordinary language and intuition leave unsettled many points about intrinsic utility, a precise account of intrinsic utility is an explication as Carnap ([1950] 1962: chap. 1) describes it. This section's account of intrinsic utility aims to assist choice, and so it respects firm intuitions for ease of application but supplements intuitions to meet decision theory's objectives. It makes intrinsic utility more precise and theoretically fruitful than are its unrefined origins.

If a proposition is realized, then it is true. The realization of a proposition representing an event is the event's occurrence. A proposition's *intrinsic utility* is the degree of intrinsic desire for its realization. This is positive in the case of an intrinsic desire, negative in the case of an intrinsic aversion, and zero in the case of an attitude of intrinsic indifference. Degree of intrinsic desire depends on quantitative comparison to a conventional unit, and assuming intrinsic indifference as a zero point forms a ratio scale for it. For the intrinsic aversions that negative degrees of intrinsic desire represent, the lower the degree, the greater the aversion. Intrinsic utility quantitatively represents intrinsic conative attitudes.

A proposition's intrinsic utility for an agent is the strength of the agent's desire that the proposition hold considering only the proposition's logical consequences. It is negative if, considering only the proposition's logical consequences, the agent has an aversion to the proposition's realization. A proposition's intrinsic utility evaluates the intrinsic features of the proposition's realization, that is, the proposition's logical consequences. Because of its narrow scope, intrinsic utility has stability as information changes.

Some introductions of utility assume that the preferences utility represents are enduring. The assumption governs preferences that are independent of information, such as basic preferences that arise from comparisons of intrinsic utilities. Binmore (1998: 362–63) distinguishes *direct utility* that attaches to final ends from *indirect utility* that attaches to means. Direct utility applies to outcomes, is insensitive to changes in information, and assesses realizations of goals. Indirect utility applies to acts, is sensitive to changes in information, and assesses prospects for realizations of goals. Intrinsic utility resembles direct utility.

Intrinsic utility registers intrinsic desires and aversions. Comprehensive utility registers both intrinsic and extrinsic desires and aversions and so differs from intrinsic utility. Whereas a proposition p's comprehensive utility $U(p)$ evaluates p's world, a proposition p's intrinsic utility $IU(p)$ evaluates p's logical consequences. Compare intrinsic utility and comprehensive utility for a typical agent. $U(\text{wealth}) > IU(\text{wealth})$ and $U(\text{smoking}) < IU(\text{smoking})$ because wealth is desired mainly as a means to other things, whereas smoking is desired mainly for pleasure and despite its adverse consequences. Letting BIT, the abbreviation for a basic intrinsic attitude, stand also for the basic intrinsic attitude's object, IU(BIT) evaluates just the logical consequences of realizing the BIT, whereas U(BIT) evaluates the world realized if the BIT were realized. Comprehensive and intrinsic utility follow different principles. Comprehensive, but not intrinsic, utility follows the expected-utility principle. Hence, U(happiness) nearly equals U(a 99 percent chance for happiness), whereas IU(happiness) is much greater than IU(a 99 percent chance for happiness). The comprehensive utility of a chance of a BIT's realization evaluates all prospects that would obtain if the chance were to obtain. The chance's intrinsic utility evaluates just what the chance entails. Given that the chance entails no BIT's realization, its intrinsic utility equals zero. A bet's comprehensive utility typically varies with information about the prospects of winning, whereas such information typically does not affect the bet's intrinsic utility.

Despite their differences, intrinsic utility and comprehensive utility agree when applied to a world. A world, a maximal proposition, entails all its features. Its intrinsic utility attends to only the features it entails, and its comprehensive utility attends to all its features. Because these features are the same, its intrinsic utility equals its comprehensive utility. For instance, IU(the actual world) = U(the actual world), given that "the actual world" abbreviates a maximal consistent proposition that represents the actual world.

Normative principles control a rational ideal agent's degrees of intrinsic desire and so intrinsic utilities as well as comprehensive utilities. Intrinsic and comprehensive utilities have structures that rationality imposes in ideal agents. A world's intrinsic and comprehensive utilities arise from realization of BITs. Intrinsic and comprehensive utilities' grounding in BITs justifies taking worlds as specifications of the BITs they realize. Any two worlds that realize the same BITs have the same intrinsic and comprehensive utilities. A full specification of the basic intrinsic attitudes that a world realizes, combined with a specification of another state the world realizes, has the same intrinsic and comprehensive utilities as the full specification of the basic intrinsic attitudes that the world realizes. Norms for rational ideal agents require these identities. This section later presents another basic norm: the intrinsic utility of a BIT's object is independent of its context of realization. The next section presents a norm of additivity. Although other norms are plausible, this chapter focuses on norms that support separability.

An agent need not assign an intrinsic utility to every proposition. A contradictory proposition lacks an intrinsic utility, and some disjunctions may lack intrinsic utilities. This section's account of intrinsic utility treats only intrinsic utilities of objects of BITs and conjunctions of such objects, including worlds represented by conjunctions of objects of BITs realized.

An analysis of a proposition's intrinsic utility for an agent at a time rests on the intrinsic utilities of objects of the agent's BITs at the time. Applications of intrinsic utility in decision theory focus on computing a world's comprehensive utility. It may be obtained from the world's intrinsic utility, which the intrinsic utilities of BITs that the world realizes generate. An application need not analyze intrinsic utilities of disjunctions or negations (except the negation of a BIT's object, which by assumption is a matter of indifference that receives zero intrinsic utility). A full treatment of intrinsic utility analyzes its application to all propositions, but a partial treatment suffices for this book's project. It needs only the intrinsic utilities of conjunctions of objects of BITs.[10]

For ranking options' worlds, any interval scale for intrinsic utility is adequate. However, for convenience, IU uses intrinsic indifference as a zero point; $IU(p) = 0$ indicates intrinsic indifference toward p. Although the function represents more information than a decision principle needs, adopting intrinsic indifference as a zero point makes the function easier to apply.

Some puzzle cases arise. Consider the intrinsic utility of a necessarily true proposition. It may seem that a rational ideal agent should be intrinsically indifferent to a necessary truth so that its intrinsic utility equals 0. However, suppose that God exists necessarily, and God's existence contributes positively to every world's utility. Then, for a rational ideal agent, it is possible that $IU(\text{God exists}) > 0$. Similarly, consider a rational ideal agent who likes the number two and intrinsically desires its existence. The number exists in

---

[10] A proposition's intrinsic utility reviews the worlds compatible with the proposition's realization. Consider a principle that requires a proposition's intrinsic utility to be the minimum intrinsic utility its realization entails; that is, the minimum among the intrinsic utilities of worlds in the set of worlds that represents the proposition (assuming that the set is an adequate representation). This principle achieves coherence between evaluations of worlds and evaluations of propositions that sets of worlds represent. However, the principle has the unintuitive consequence that a necessary proposition has the lowest possible intrinsic utility. The set of all worlds represents it, and some world maximizes intrinsic aversion and minimizes intrinsic desire. Also, according to the principle, the intrinsic utility of realizing a basic intrinsic desire is negative if its realization is compatible with realizing a stronger basic intrinsic aversion. According to a rival principle, a proposition's intrinsic utility equals the minimum contribution the proposition's realization makes to a world that realizes it. Unfortunately, this principle requires a disjunction of the objects of a basic intrinsic desire and a basic intrinsic aversion to have the same intrinsic utility as the basic intrinsic aversion's object. Chisholm (1975: 302–3) and Weirich (2009) advance two-account principles that allow the disjunction to have a different intrinsic utility. This chapter's characterization of intrinsic utility accommodates an agent's not assigning any intrinsic utility to the disjunction.

every world, and its existence contributes positively to each world's utility so that $IU(2 \text{ exists}) > 0$. A structural account of propositions resolves the puzzle. It is possible that $IU(p \vee \sim p) = 0$ although $IU(\text{God exists}) > 0$ and $IU(2 \text{ exists}) > 0$, assuming that a proposition is not a set of possible worlds but rather has structure so that necessary propositions may differ; the structural account of propositions prevents the utility assignment's incon-sistency, granting that not every proposition entails every necessary truth.

Consider the intrinsic utility of the negation of a BIT's object. It may seem that for a rational ideal agent, $IU(\text{BIT}) = -IU(\sim\text{BIT})$. However, a rational ideal agent may be intrinsically indifferent to the nonrealization of a simple pleasure rather than intrinsically averse to its nonrealization. The generalization does not hold in all cases.

BITs are such that each may be realized alone because no BIT's object entails any other BIT's object. The intrinsic utility of a BIT's realization equals the intrinsic utility of the world in which that BIT is the only one realized. Intrinsic attitudes toward worlds yield intrinsic attitudes toward BITs, and ordinary comprehensive desires toward worlds yield intrinsic attitudes toward worlds. Comprehensive utilities of worlds therefore yield the intrinsic utilities of objects of BITs. For a BIT, $IU(\text{BIT})$ equals the comprehensive utility of the world realizing only the BIT. Hence, one may infer $IU(\text{BIT})$ from $U(w)$ if w is a world that realizes only the BIT.

For a BIT, the quantity $IU(\text{BIT})$ is the same in all worlds that realize the BIT. This constancy is a norm rather than a stipulation. Instead of defining intrinsic utility and BITs so that BITs make a constant contri-bution of intrinsic utility to worlds that realize them, this section uses the independence of reasons for BITs to derive normatively their constant contribution.[11]

A proposition's intrinsic utility may put aside context, including infor-mation, because intrinsic utility is independent of context. A proposition's

---

[11] Suppose that the intrinsic utility of a world has been defined. Consider a definition of a proposition's intrinsic utility that takes it as the constant contribution the proposition makes to the intrinsic utility of a world that realizes it. This definition does not reach all propositions. Imagine a disjunction of the objects of two basic intrinsic desires of unequal intensity. The constant-contribution interpretation assigns no intrinsic utility because the disjunction does not make a constant contribution. The constant-contribution interpretation limits intrinsic utility to propositions that make a constant contribution to a world's intrinsic utility. Chisholm (1975: 295), Gibbard ([1972] 1990: 145–50), and Zimmerman (2001: chap. 5) consider a constant-contribution interpretation of intrinsic value. Assuming this interpretation, Zimmerman declares "evaluatively inadequate" a state of affairs speci-fied by a disjunction of a basic good and a basic evil. The state of affairs does not make a constant contribution to the intrinsic value of worlds that realize it and so lacks an intrinsic value. Zimmerman's restriction of intrinsic value skirts difficulties with disjunctions that Quinn (1974) describes and Chisholm (1975) attempts to address.

logical consequences, the elements of intrinsic utility's evaluation of the proposition, are the same in all contexts. Moreover, the reasons for distinct BITs are independent. No BIT is a reason for another. Intrinsic conative attitudes do not ground BITs. Hence, the intrinsic utilities of their objects are independent of suppositions about circumstances. A norm for ideal agents ensures this independence.[12] If an ideal agent has a basic intrinsic desire that p, then she should with the same intensity intrinsically desire that p whether or not q holds for any q not entailing p. A similar norm holds for basic intrinsic aversions. Hence, BITs make a constant contribution to intrinsic utilities. In an ideal agent, rationality requires their realizations to make a constant contribution to combinations that realize them. The norm applies to a single BIT's realization and also to realization of a conjunction that represents realization of a set of BITs.

The *marginal intrinsic utility* of a proposition q given a proposition p, $IU(q \mid p)$, is an evaluation of the difference between realization of p and realization of $(p \ \& \ q)$, considering the propositions' implications. For example, the marginal intrinsic utility of wisdom given health evaluates the difference between health and its combination with wisdom, considering the implications of health and also the implications of its combination with wisdom. The evaluation assesses a change from realization of p to realization of $(p \ \& \ q)$ and does not assess a single proposition's realization. $IU(q \mid p)$ is q's contribution to the intrinsic utility of $(p \ \& \ q)$ given that p.

The implications of q are the same whether or not p holds, but their evaluation may change assuming that p holds if it assesses their contribution to intrinsic utility, as marginal intrinsic utility does. An agent supposes that p when assigning q's marginal intrinsic utility given p. The assignment depends on the agent's appraisal of q's contribution to intrinsic utility under that supposition.[13]

Consider a composite. A component's intrinsic utility in isolation, its marginal intrinsic utility when taken as first in the composite, is the composite's intrinsic utility when other components contribute no intrinsic utility. In a composite with two ordered components, an assessment of the second component's marginal intrinsic utility compares two composites. One composite has the first component and a substitute for the second component that makes no contribution, and the other composite has the

---

[12] Norms also move from reasons to a judgment of independence, not just the fact of independence.

[13] For some types of conditional intrinsic utility, the intrinsic utility of q given p equals the intrinsic utility of $(p \ \& \ q)$. This is so if the type of conditional intrinsic utility takes the implications of q given p to be the same as the implications of $(p \ \& \ q)$. Marginal intrinsic utility takes the implications of q given p to be just q's implications.

first component and the second component. The comparison uses the first component's intrinsic utility in isolation and the second component's intrinsic utility given the first component.

For a rational ideal agent, $IU(p \& q)$ exists only if p and q are compatible. Assume that the two propositions are compatible. An ideal agent, if rational, assigns to $IU(q \mid p)$ the difference between $IU(p \& q)$ and $IU(p)$, given that these quantities both exist. That is, $IU(q \mid p) = IU(p \& q) - IU(p)$. The equation does not hold simply because of the definition of marginal intrinsic utility but also because of rationality's requirement that an ideal agent's evaluation of the difference between the implications of $(p \& q)$ and of q equal the difference between her intrinsic desires that $(p \& q)$ and that q. Thus, IU(wisdom | health), the intrinsic utility that wisdom contributes given health, equals the difference between the intrinsic utility of wisdom conjoined with health and the intrinsic utility of health. Intuition supports the norm, in fact, so well that compliance seems to follow from the definition of marginal intrinsic utility alone. However, the norm adopts a convention that goes beyond the definition. The convention evaluates the difference between the realizations of two propositions using the difference of their intrinsic utilities rather than, for example, their ratio. It uses the difference in the quantities to represent an evaluation of the difference in the propositions' realizations.[14]

The marginal intrinsic utility of a proposition is relative to a proposition. Hence, in a conjunction of propositions, the marginal intrinsic utility of a conjunct taken alone, that is, relative to a logical truth, may differ from its marginal intrinsic utility relative to the other conjunct. $IU(q)$ may differ from $IU(q \mid p)$. The proposition q is *contributionally independent* of the proposition p if and only if $IU(q)$ equals $IU(q \mid p)$. In that case, p's realization does not affect the marginal intrinsic utility of q's realization. Changing the condition for q's marginal intrinsic utility from a logical truth to p has no effect on q's marginal intrinsic utility. If in the combination of p's realization and q's realization, q's realization comes first, its marginal intrinsic utility (taken with respect to a logical truth) is $IU(q)$. If its realization comes after p's realization, its marginal intrinsic utility is $IU(q \mid p)$. Given q's independence from p, $IU(q) = IU(q \mid p)$. It follows that the order of the two propositions' realizations does not affect q's marginal intrinsic utility. Independence for pairs of propositions is a type of noncomplementarity.

---

[14] Velleman (2000) defines marginal value as a difference in values. In contrast with marginal intrinsic utility, marginal value's equality with a difference in values is not normative.

A proposition q's constant contribution to intrinsic utility follows from q's independence of every proposition p with which it is compatible.[15]

Consider the marginal intrinsic utility of realizing a BIT given a proposition p. Using "BIT" to stand also for the BIT's object, $IU(\text{BIT} \mid p)$ expresses this quantity. It is an evaluation of the difference between realization of p and realization of $(p \ \& \ \text{BIT})$, considering the propositions' implications. A rational ideal agent assigns to $IU(\text{BIT} \mid p)$ the difference between $IU(p \ \& \ \text{BIT})$ and $IU(p)$. Given that realization of a BIT is contributionally independent of p, the difference equals $IU(\text{BIT})$. Suppose that an agent has a basic intrinsic desire to be wise, and being wise is contributionally independent of being tall, then the marginal intrinsic utility of being wise given tallness equals the intrinsic utility of being wise. Abbreviating, $IU(\text{wise} \mid \text{tall}) = IU(\text{wise})$.

Because the reasons for a BIT are independent of all other intrinsic attitudes, an ideal agent, if rational, assigns the same intrinsic utility to the BIT's object given any compatible proposition that does not entail the BIT's object, provided that the conjunction of the two propositions is *nonproductive*, that is, entails only BITs that some conjunct entails. A BIT's special features and rationality's constraints on an ideal agent's intrinsic attitudes make the intrinsic utility of a BIT's realization contributionally independent of the context of its realization. That is, the marginal intrinsic utility of a BIT's object is the same given any compatible proposition p not entailing its object, provided that the conjunction of the two propositions is nonproductive. For a BIT and all p meeting the conditions described, $IU(\text{BIT} \mid p) = IU(\text{BIT})$.[16] A BIT's realization makes a constant contribution to the intrinsic utility of combinations that realize it. This type of independence of BITs from the context of their realizations is normative and not simply a matter of definition. The norm has intuition's support because of the independence of the reasons for a BIT. Section 3.5 uses the constancy of a BIT's contribution to argue for intrinsic utility's additivity.[17]

---

[15] It may seem that $IU(q \mid p) = IU(p \ \& \ q)$ when q is independent of p because a ranking of independent q's according to intrinsic utility given p is the same as a ranking of their conjunctions with p according to intrinsic utility. However, if $IU(q \mid p) = IU(p \ \& \ q)$ given q's independence of p, then $IU(p \ \& \ q)$ is constant for all q independent of p. This is implausible. Instead, $IU(q \mid p) = IU(p \ \& \ q)$ if the intrinsic attitude toward p is indifference.

[16] Suppose that because the conjunction $(p \ \& \ q)$ is productive of r's realization, $IU(p \ \& \ q) = IU(p) + IU(q) + IU(r)$. Then, $IU(q \mid p) = IU(p \ \& \ q) - IU(p) \neq IU(q)$, so independence fails if $IU(r) \neq 0$. Also, if p entails q, then $IU(p \ \& \ q) = IU(p)$, so $IU(q \mid p) = IU(p \ \& \ q) - IU(p) = 0$. In this case, independence fails if $IU(q) \neq 0$.

[17] Buchak (2013) defines independence of goods so that it implies an equality of differences in utilities and so an equality of utilities and marginal utilities. According to her definition, goods A and B are

3.4   **Intrinsic-utility analysis**

Expected-utility analysis spreads an option's utility along the dimension of chances for possible outcomes. It takes the chances as parts of the option's realization. Intrinsic-utility analysis spreads a world's utility along the dimension of basic intrinsic attitudes (BITs). It takes the world's realizations of BITs as parts of the world. The two forms of analysis operate together when information is incomplete. For simplicity, this section's formulation of intrinsic-utility analysis assumes certainty of an option's world. Section 3.8 adds expected-utility analysis to handle uncertainty.[18]

Contemporary utility theorists generally do not pursue the traditional project of explaining an option's utility using the goals the option's realization meets or frustrates. This project requires analyzing the utility of an outcome, that is, a possible world, using the world's features. Complementarity among a world's features seems to thwart the analysis. Utilities of worlds appear to be fundamental. Some theorists however explore ways of overcoming the problems.

Keeney and Raiffa ([1976] 1993) and Baron (2008: chap. 14) present a method of analyzing an outcome according to objectives it realizes. The method is *multi-attribute–utility* analysis. It divides an option's outcome into realizations of various objectives and computes the outcome's utility using the utilities of realizing the objectives. *Intrinsic-utility analysis*, a version of multi-attribute–utility analysis that Weirich (2001: chap. 2) introduces, takes an agent's objectives as realizations of basic intrinsic

---

independent if and only if, using A and B to stand for possession of the corresponding goods, $U(A \& B) - U(A \& {\sim}B) = U({\sim}A \& B) - U({\sim}A \& {\sim}B)$. The utility difference between having B and not having B is the same whether or not one has A. That is, B's marginal utility is independent of possession of A. If $U({\sim}A \& {\sim}B) = 0$, this type of independence implies $U(A \& {\sim}B) + U({\sim}A \& B) = U(A \& B)$. The utility of A without B plus the utility of B without A equals the utility of A and B. Independence as defined thus implies a type of additivity. This type of additivity governs comprehensive utility and is not a norm unless the type of independence is a norm. Buchak does not argue for independence of goods but just assumes it as an empirical fact in some cases. Two BITs may fail to be independent in the required sense because of information's effect on comprehensive utility. Suppose that A and B are objects of basic intrinsic desires. Also, suppose that $U({\sim}A \& {\sim}B)$ and $U({\sim}A \& B)$ both equal 0 because $U({\sim}A \& {\sim}B)$ sets the utility scale's zero point and because given ${\sim}A$ having B brings with certainty a counterbalancing bad. Imagine that $U(A \& {\sim}B)$ is positive but that $U(A \& B)$ is greater because having both A and B is better than having just A. Then, the type of additivity fails, so A and B are not independent goods according to the definition. This chapter makes normative the equality of a marginal intrinsic utility and a difference of intrinsic utilities and also makes normative the contributional independence of realizations of BITs. Its support for intrinsic utility's additivity, drawing on these principles, is normative rather than definitional.

[18] I thank Xiaofei Liu for bibliographic research for this section and for helpful comments. For many suggestions, I also thank Marion Ledwig, Matt McGrath, Alec Walen, and the audiences for presentations at the 2003 meeting of the Society for Exact Philosophy, the 2003 conference on Logic, Methodology, and Philosophy of Science, and the 2004 Eastern APA meeting.

desires and nonrealizations of basic intrinsic aversions. It takes an option's outcome as the option's world and divides the world's utility into the intrinsic utilities of the world's realizations of BITs. Intrinsic-utility analysis assists utility-maximizing decision procedures by calculating comprehensive utilities of options' outcomes.[19]

Decision theory's traditional principle of adding pros and cons is imprecise, and so as a substitute this chapter formulates a precise version using its account of intrinsic utility. Its addition principle for intrinsic utilities of BITs' objects is an individualistic analog of the utilitarian addition principle for social utility. The analogy takes the intrinsic utility of realizing a BIT as the analog of the personal utility of realizing a proposition. It takes summing intrinsic utilities of objects of BITs a world realizes to calculate the intrinsic utility of the world's realization as the analog of summing the personal utilities of a proposition's realization to calculate the social utility of the proposition's realization. The independence of the utilities added

---

[19] Nelson (1999) describes multi-attribute–utility models designed to predict consumer choices among brands, such as brands of cars, using brand attributes, such as horsepower, miles per gallon, comfort, and style. Intrinsic-utility analysis is normative in contrast with the empirical models Nelson describes.

Besides multi-attribute–utility analysis, another predecessor of intrinsic-utility analysis is "reason-based choice." Shafir, Simonson, and Tversky (1993) investigate how reasons, such as features of options making them attractive or unattractive, influence choices. History and law typically explain choices under the assumption that choices rest on reasons, whereas economics typically explains choices under the assumption that they comply with expected-utility theory. Shafir et al.'s account of reason-based choice is descriptive rather than normative, informal rather than formal, and supplements rather than replaces expected-utility theory. Their account's attention to reasons resembles intrinsic-utility analysis's attention to features of options that direct evaluation. However, intrinsic-utility analysis is precise, normative, and an articulation of expected-utility theory that uses choices to infer, rather than define, probabilities and utilities.

Following a suggestion of Pettit (2002: 167) to use utilities of properties a world instantiates to obtain its utility, Dietrich and List (2013) present a reason-based method of rationalizing an agent's choice behavior. Their method explains a choice in a decision problem using "motivationally salient" properties of the problem and the options it offers. For example, a shopper selecting yogurt at a supermarket may attend to whether a type of yogurt is fruit flavored, low fat, free of artificial sweeteners, and the cheapest on display but may ignore the place of its production and its position in the display. Although, as does intrinsic-utility analysis, reason-based rationalization examines an outcome's features to explain its overall evaluation, in contrast with intrinsic-utility analysis, it does not maintain that certain properties of an outcome are separable considerations that a simplified evaluation of the outcome may omit.

A related method of evaluation for a commodity uses a hedonic price, that is, a price suggested by the commodity's features. As Rosen (1974) explains, in a model for hedonic prices, a vector of characteristics represents a commodity and settles its hedonic price, which equals its observed price in a market equilibrium. The model assumes that consumers and producers optimize, given a ranking of outcomes. Malpezzi (2003) reviews hedonic-pricing models using characteristics of a commodity to predict or explain its price. For example, a model of housing prices may use square footage and location, among other characteristics, to predict a house's price. Intrinsic-utility analysis similarly uses an option's features to evaluate the option, but the analysis is normative, for an individual, and uses separable features so that the option's efficient evaluation may put aside some features.

grounds both addition principles. This section just presents the addition principle for intrinsic utilities. The next section establishes intrinsic utility's additivity using BITs' independence.

A nonmaximal proposition carries less information than does a maximal proposition that represents a world, and in this sense it is a smaller proposition. It describes only a portion of a world. People apparently have BITs toward propositional objects smaller than worlds and form intrinsic utilities for conjunctions of BITs' objects additively in at least some cases. Take preferences for collections of music in an iPod. Suppose that Sally likes song *A* twice as much as song *B* and likes song *B* twice as much as song *C*. Then, she may like the collection {*A*, *B*} twice as much as the collection {*B*, *C*}, as additivity requires assuming that desires to have the songs in her iPod are BITs. The structure of people's intrinsic attitudes, and the extent to which they comply with normative principles for intrinsic attitudes, are empirical matters. This chapter makes no empirical assumptions about intrinsic attitudes but instead works within a normative model that simply stipulates that an agent has BITs that generate intrinsic utilities of worlds. The model treats an ideal agent whose BITs toward nonmaximal propositions (that parts of worlds realize) are effective reasons for the agent's intrinsic attitudes to maximal propositions (that represent worlds) and are complete reasons for all the agent's intrinsic attitudes and so are reasons for their intensities as well their contents.

Under the chapter's idealizations about agents, a possible world's intrinsic utility is the sum of the intrinsic utilities of the objects of BITs the world realizes. Using BIT_i to stand for the basic intrinsic attitude's object, the addition principle for worlds states that

$$IU(w) = \sum{}_i IU(BIT_i)$$

where BIT_i ranges over the objects of the BITs that *w* realizes. If an agent's only BITs are for health and wisdom, so that health and wisdom constitute a world, and *IU*(health) is twice *IU*(wisdom), which serves as the unit for *IU*, then *IU*(health and wisdom) = 2 + 1 = 3. If a world entails no BIT's realization, it is an object of intrinsic indifference, and its intrinsic utility is zero.[20]

Intrinsic utility is additive because realizing a BIT has the same intrinsic utility no matter which other BITs are realized. As this chapter defines a BIT, the reasons for a BIT are independent of the context of its realization,

---

[20] In moral theory, Gibbard ([1972] 1990: 145–50) and Zimmerman (2001: chap. 5) advance an addition principle for a world's intrinsic value.

and so its realization has the same intrinsic utility in any context. Realizing basic intrinsic desires does not have diminishing marginal intrinsic utility. Unlike satisfying desires for potato chips, satiation does not reduce the intrinsic utility of realizing a basic intrinsic desire. Suppose that realizing basic intrinsic desires is productive in the sense that realizing many basic intrinsic desires entails realization of another BIT, an aversion to excess. Then, adding to realization of some basic intrinsic desires realization of another basic intrinsic desire BIT_n may increase total intrinsic utility by less than the intrinsic utility of realizing BIT_n. This creates the impression of diminishing marginal intrinsic utility. However, the realization of BIT_n with other basic intrinsic desires entails the realization of the basic intrinsic aversion to excess, whose realization lowers total intrinsic utility. An evaluation of realizations of all BITs, including the basic intrinsic aversion to excess, eliminates the impression of diminishing marginal intrinsic utility.

The formula for a world's intrinsic utility asserts the division of the intrinsic utility of a whole into the intrinsic utilities of its parts, taking the whole as a world and its parts as the objects of the BITs that the world realizes. Taking the world to specify the BITs realized, it is a conjunction, and each BIT's realization forms a conjunct. The intrinsic utility of the conjunction is the sum of the intrinsic utilities of these conjuncts. A world's intrinsic utility comes from the BITs it realizes because they implicitly count all the intrinsic attitudes the world realizes. The sum counts the nonbasic intrinsic attitudes realized by counting the realizations of the BITs that are the agent's reasons for the nonbasic attitudes. The conjunction of the objects of the BITs realized in a world entails the realizations of the intrinsic attitudes realized there. Hence, focusing on the conjunction of the objects of the BITs realized omits no relevant considerations. The conjunction's logical consequences include everything that matters to the world's intrinsic utility. Also, counting only BITs' realizations prevents double counting intrinsic attitudes whose realizations other intrinsic attitudes' realizations entail. Therefore, summation of intrinsic utilities of the BITs' objects neither omits nor double counts any relevant consideration. Summing their intrinsic utilities yields the world's intrinsic utility just as summing the weights of atoms yields an object's weight. By summing atoms' weights, one takes account of the weights of molecules and does not count twice any object's weight. Similarly, by counting realizations of BITs, one counts realizations of all relevant intrinsic attitudes and counts none twice.

Intrinsic-utility analysis breaks down the intrinsic utility of a possible world, taken as a maximal consistent proposition, into the intrinsic utilities of the objects of the basic intrinsic attitudes the world realizes. The weights of bricks in a wall constitute rather than cause the wall's weight. In contrast, because for a rational ideal agent reasons are causes, the intrinsic utilities of a world's realizations of BITs cause rather than constitute the world's intrinsic utility. Because of principles of rationality, a world's intrinsic utility sums numeric evaluations of the world's realizations of BITs. Additivity is a norm for degrees of intrinsic desire and not a definitional truth about degrees of intrinsic desire.

Adding utilities of BITs a world realizes to calculate the world's intrinsic utility assumes that for a rational ideal agent nonrealization of a BIT may be ignored. Suppose that a BIT's nonrealization does not entail any BIT's realization. Then, it is a matter of intrinsic indifference. Suppose that its nonrealization is not a matter of indifference because it entails another BIT's realization. Perhaps a runner has a basic intrinsic desire to win a race and a basic intrinsic aversion to losing the race, an intrinsic aversion that the intrinsic desire does not ground. Then, for each BIT the intrinsic utility of the other BIT's realization takes account of the BIT's nonrealization. If this point is wrong, the addition principle restricts itself to cases in which a BIT's nonrealization is a matter of indifference. In any case, in a conjunction representing a world, the intrinsic utility of a conjunct specifying that a BIT is not realized is zero, and so the summation for the conjunction ignores it. A world's specification may just say which BITs it realizes, taking as understood that BITs not mentioned are not realized. BITs a world does not realize do not affect its intrinsic utility, so a proposition specifying the BITs that the world realizes represents it. Its propositional representation may omit specifying nonrealization of BITs. The intrinsic utility of a world, a proposition specifying for each BIT whether it is realized, equals the intrinsic utility of a conjunction of the objects of BITs that the world realizes.

A world's canonical representation for an agent specifies every feature of the world that the agent cares about. Intrinsic-utility analysis reduces the representation first by focusing on realizations and nonrealizations of BITs and then by focusing on realizations of BITs. It assumes that specifying a world's realization of BITs adequately represents the world for calculating its intrinsic utility.[21]

An extension of intrinsic-utility analysis applies to a nonmaximal consistent conjunction of BITs' objects. The conjunction may entail a

---

[21] Hansson (2001: 58) distinguishes holistic and aggregative approaches to preferences among worlds (or intrinsic desires concerning worlds) and describes assumptions of the aggregative approach.

BIT's object that is not a conjunct. Hence, the intrinsic utility of realizing a pair of BITs may not equal the sum of the intrinsic utilities of realizing the BITs. In some cases, realizing a pair of BITs may entail realizing a third BIT. Then, the intrinsic utility of realizing the pair is the sum of the intrinsic utilities of all objects of BITs that the pair's realization entails. Consider the equality $IU(BIT_1 \,\&\, BIT_2) = IU(BIT_1) + IU(BIT_2)$. This equality may fail if BIT_1 realized together with BIT_2 entails another BIT's realization. For example, BIT_1 and BIT_2 may be basic intrinsic desires for levels of pleasure during, for BIT_1, a certain temporal interval and, for BIT_2, an immediately succeeding temporal interval. Suppose that the levels of pleasure are the same so that joint realization of BIT_1 and BIT_2 entails realization of BIT_3, a basic intrinsic desire for a constant level of pleasure during the combination of the two temporal intervals. Given that the realization of BIT_1 and BIT_2 entails the realization of BIT_3, $IU(BIT_1 \,\&\, BIT_2) = IU(BIT_1 \,\&\, BIT_2 \,\&\, BIT_3) = IU(BIT_1) + IU(BIT_2) + IU(BIT_3)$. An analysis of $IU(BIT_1 \,\&\, BIT_2)$ must consider not only the BITs whose objects are conjuncts but also BIT_3.

The extended addition principle states that the intrinsic utility of a consistent conjunction of objects of BITs is the sum of the intrinsic utilities of the objects of BITs whose realizations the conjunction entails. Let $\&_i BIT_i$ conjoin objects of BITs that i indexes.[22] $IU(\&_i BIT_i) = \sum_j IU(BIT_j)$, where BIT_j ranges over objects of BITs that the conjunction $\&_i BIT_i$ entails. The summation includes just once each BIT's object that the conjunction entails.[23]

The stipulation about consistency puts aside, for example, the conjunction of two objects of basic intrinsic desire from a family of mutually exclusive objects; say, pleasures of different degrees at a single time. Inconsistent conjunctions lack intrinsic utilities. The restriction to conjunctions whose conjuncts are objects of BITs cannot be weakened to allow conjuncts that are objects of intrinsic attitudes in general. It need not be the case that $IU([x \,\&\, y] \,\&\, [y \,\&\, z]) = IU(x \,\&\, y) + IU(y \,\&\, z)$, where x, y, and z are objects of BITs, and therefore $[x \,\&\, y]$ and $[y \,\&\, z]$ are objects of intrinsic attitudes but not objects of BITs. The sum double counts y.

The addition principle for worlds is a corollary of the general addition principle for conjunctions. Although a realization of a conjunction of BITs' objects may entail realization of other BITs, realization of a conjunction that represents a world does not entail realization of any BIT whose object is

---

[22] A vacuous conjunction that entails no BIT's realization is an object of intrinsic indifference, and its intrinsic utility is zero.

[23] Carlson (2001) presents a similar addition principle that he attributes to G. E. Moore.

not a conjunct because the proposition representing the world conjoins the objects of all BITs that the world realizes. Adding its conjuncts' intrinsic utilities therefore yields the conjunction's intrinsic utility according to the principle for conjunctions. The principle for worlds follows because the world's intrinsic utility equals the conjunction's intrinsic utility.

The addition principle for conjunctions does not generalize for all propositions. It is not true for every proposition p that $IU(p)$ equals the sum of the intrinsic utilities of the objects of BITs whose realization p entails. In particular, the generalization does not hold for disjunctions. Suppose, for example, that x and y are mutually exclusive objects of basic intrinsic desire and are the only objects of BITs. Consider $IU(x \lor y)$. By the principle for all propositions, $IU(x \lor y) = 0$ because the disjunction does not entail satisfaction of any BIT. However, a rational ideal agent may attribute no intrinsic utility to the disjunction. The principle for all propositions fails given that the intrinsic utility of the disjunction does not exist. Also, if realizations of the basic intrinsic desires have the same positive intrinsic utility, a rational ideal agent may attribute that intrinsic utility, not zero, to their disjunction. Again, the principle for all propositions fails.

The restriction on additivity does not count against this chapter's interpretation of intrinsic utility because intuition does not support unrestricted additivity. Intuition does not expect the intrinsic utility of a disjunction of objects of BITs to be the sum of the intrinsic utilities of the BITs whose realizations the disjunction entails. The addition principle applies to a conjunction of objects of BITs because the relevant logical consequences of such a conjunction are the BITs whose realization it entails. Intuition recognizes that this point about relevant logical consequences does not apply to all propositions.

3.5 Additivity's support

This section establishes a norm for degrees of intrinsic desires: for an ideal agent, rationality requires that the degree of intrinsic desire for realization of a conjunction of objects of BITs equal the sum of the degrees of intrinsic desire for realizations of objects of BITs that the conjunction entails. Intrinsic utilities are a rational ideal agent's degrees of intrinsic desire. So, norms for an ideal agent's degrees of intrinsic desire yield properties and relations of intrinsic utilities. Because additivity is a norm for degrees of intrinsic desire, intrinsic utilities are additive. In an ideal agent, rationality ensures additivity of intrinsic utilities, taken as degrees of intrinsic desire, as it ensures additivity of probabilities, taken as degrees of belief.

Consider an ideal agent who has a basic intrinsic desire for an hour at the beach today and a twice as great basic intrinsic desire for two hours at the beach tomorrow. Using the first desire as a unit, it seems that the agent's intrinsic desire for realizing both desires should be one plus two, or three units. The norm of additivity has intuition's backing in such cases. An ideal agent's cognitive abilities remove excuses for noncompliance.

The normative consequences of additivity's normativity also hold. Intrinsic utilities have the relations that follow from their additivity. For a conjunction of objects of BITs, intrinsic utility is constant under commutation and reassociation. For example, $IU(BIT_1 \& BIT_2) = IU(BIT_2 \& BIT_1)$, and $IU((BIT_1 \& BIT_2) \& BIT_3) = IU(BIT_1 \& (BIT_2 \& BIT_3))$. Moreover, suppose that a conjunction of objects of BITs entails only objects of BITs that are conjuncts, so that the conjunction is *nonproductive*. Then, being a matter of basic intrinsic indifference makes a conjunct an identity element. Given that the object of BIT_2 is a matter of basic intrinsic indifference, $IU(BIT_1 \& BIT_2) = IU(BIT_1)$, for example.

Because intrinsic utilities are negative for basic intrinsic aversions, intrinsic utilities of objects of BITs and their conjunctions have addition's property that $x + (-x) = 0$, the property of opposites, and its property that $(-x) + (-y) = -(x + y)$, the property of the opposite of a sum. An intrinsic attitude's *opposite* is a contrary intrinsic attitude of the same intensity. For intrinsic utilities, according to the property of opposites, zero equals the intrinsic utility of the nonproductive conjunction of a BIT's realization with an opposite BIT's realization. So, if a basic intrinsic desire for a pleasure has the same intensity as a basic intrinsic aversion to a pain, the intrinsic utility of the realization of both attitudes, assuming that it is nonproductive, equals zero. According to the property of the opposite of a sum, the intrinsic utility of the nonproductive conjunction of the objects of two basic intrinsic aversions equals the negative of the intrinsic utility of the nonproductive conjunction of two basic intrinsic desires that are, respectively, opposite to the basic intrinsic aversions. So, if two basic intrinsic desires for pleasures have the same intensities, respectively, as two basic intrinsic aversions to pains, then the sum of the intrinsic utilities of realizations of the aversions equals the negative of the sum of the intrinsic utilities of the realization of the desires, assuming that neither combination is productive.

Furthermore, the field of conjunction for objects of BITs includes all objects of BITs and is closed under a member's conjunction with an object of a BIT not already entailed by objects of BITs to which the BIT is conjoined. For conjunctions of objects of BITs, intrinsic utility's applications to conjuncts are comparable (being on the same scale for intrinsic desire), its application to a conjunct is independent of its application to another conjunct, and the contribution its application to a conjunct makes to its application to a conjunction is independent of the contribution of its application to other conjuncts, so that the conjunct's location is separable from the other conjuncts' locations, as Section 3.7 explains.

That intrinsic utility has properties that its additivity entails supports intrinsic utility's additivity. However, this section's main argument for the principle of addition uses BITs' independence.[24] The argument assumes a quantitative representation of intrinsic attitudes that yields degrees of intrinsic desire. The argument does not define degrees of intrinsic desire as representations of intrinsic preferences among propositions. Degrees of intrinsic desires may generate intrinsic preferences so that the properties of intrinsic preferences derive from the properties of degrees of intrinsic desire. The representation of intensities of intrinsic desires and aversions does not assume additivity, in contrast with standard methods of measurement that Section 3.9 reviews.

Intrinsic utility's additivity is partly conventional. Given some representations of intensities of intrinsic desires and aversions, the intrinsic utility of a nonproductive conjunction of BITs' objects may not equal the sum of the conjuncts' intrinsic utilities. Additivity assumes a congenial representation of intensities of intrinsic desires and aversions, taking, for instance, a proposition of basic intrinsic indifference as an identity element. Conventions represent degrees of intrinsic desire so that when they are fully rational in an ideal agent, they are additive.

Not all separable functions may represent intrinsic utility's separability. The function representing its separability must simultaneously represent ways of increasing and decreasing the intrinsic utility of a nonproductive conjunction of BITs' objects. Multiplication, although a separable function, fails to represent increases and decreases. Conjoining a basic intrinsic desire's object to a nonproductive conjunction of BITs' objects to produce another nonproductive conjunction generates a conjunction with greater intrinsic utility than the original conjunction. However, the product of the intrinsic utility of the original conjunction and the intrinsic utility of the desire's object is less than the original conjunction's intrinsic utility if the intrinsic utility of the desire's object is smaller than the unit. The product is negative if a basic intrinsic aversion's object replaces the desire's object, and the original conjunction's intrinsic utility is positive, no matter how small

---

[24] Weirich (2001: chap. 2) uses assumptions about causal relations among the grounds of intrinsic utilities to argue for additivity. This section's argument for additivity uses weaker assumptions.

the intrinsic utility of the aversion's object and how great the original conjunction's intrinsic utility. The product is zero if an object of basic intrinsic indifference replaces the desire's object. Addition, but not multiplication, represents the effect on intrinsic utility of nonproductively conjoining a BIT's object to a nonproductive conjunction of BITs' objects.[25]

Recall from Section 3.3 that $IU(q \mid p)$ is the marginal intrinsic utility of q given p, that is, the intrinsic utility that q's realization contributes given p's realization. This section uses Section 3.3's norm of marginal intrinsic utility and norm of independence for objects of BITs to establish additivity, first for nonproductive conjunctions and then for all conjunctions of objects of BITs.

Given a congenial representation of intensities of intrinsic attitudes, the norm of marginal intrinsic utility yields additivity for two independent propositions with intrinsic utilities. According to the norm of marginal intrinsic utility: $IU(q \mid p) = IU(p \& q) - IU(p)$. If q is independent of p, then $IU(q \mid p) = IU(q)$. Hence, $IU(q) = IU(p \& q) - IU(p)$. Therefore, $IU(p \& q) = IU(p) + IU(q)$. Intrinsic utility is additive given independence.

The norm of independence requires that the marginal intrinsic utility of a BIT's realization given a proposition p meeting some conditions equal the intrinsic utility of the BIT's realization. According to the norm, $IU(\text{BIT} \mid p) = IU(\text{BIT})$ if p does not entail the BIT's object, and conjunction of p and the BIT's object is nonproductive. Although no BIT's object entails another BIT's object, independence defined in terms of marginal contribution does not entail additivity, just constant marginal contribution of some type. The independence of BITs supports but does not entail that $IU(\text{BIT}_1 \& \text{BIT}_2) = IU(\text{BIT}_1) + IU(\text{BIT}_2)$. Independent contribution is not the same as additivity because it leaves open the manner of contribution. It is compatible with a compositional function that is multiplicative rather than additive. However, the norm of marginal intrinsic utility, which adopts additive aggregation, and the norm of independence together yield additivity for a nonproductive conjunction of two objects of BITs.

BITs are grounded independently of one another (although their objects may be exclusive and so not logically independent). Consequently, their

---

[25] This paragraph compares an n-tuple of values of variables with an $(n+1)$-tuple of values of variables. Separability applies to an n-tuple of variables. To compare only n-tuples of values of variables, a default value for a variable may replace addition of the variable's value. For a multiplicative representation of the n-tuples' order, the default value must be 1. However, this default value cannot be intrinsic utility's unit because an object of intrinsic indifference receives the default value. An additive representation permits using 0 as the default value and using an object of intrinsic desire to set intrinsic utility's unit.

realizations have independent effects on the intrinsic utilities of their realizations' combinations. The independence of BITs justifies addition of intrinsic utilities of their objects to obtain the intrinsic utilities of conjunctions of their objects. The intrinsic utility of a BIT's object is an independent contribution to the intrinsic utility of a conjunction of BITs' objects that entails the BIT. The intrinsic utility's contribution is independent, although a BIT's realization may, given some BITs' realizations, produce another BIT's realization. Additivity holds as long as it registers all BITs realized. The principle of addition works because for a BIT, its intrinsic utility is its constant contribution to combinations of BITs' realizations. Its contribution is constant even if its realization may combine with another BIT's realization to entail a third BIT's realization.

The combination of two propositions' realizations may have an order that their conjunction does not represent because the order of conjuncts is arbitrary. A combination's intrinsic utility is a sum of its elements' marginal intrinsic utilities according to the elements' order. For example, $IU(p \& q) = IU(p) + IU(q \mid p)$, assuming that p comes first in the combination of p and q so that $IU(p)$ equals p's marginal intrinsic utility. If q is first in the combination, then $IU(q)$ equals q's marginal utility, and the equation for the combination is $IU(p \& q) = IU(q) + IU(p \mid q)$. Comparing $IU(q)$ and $IU(q \mid p)$ compares q's marginal utility when it comes first and when it comes second. Possibly, $IU(q) \neq IU(q \mid p)$. Then, q's marginal utility depends on its position. If q is independent of p, then $IU(q) = IU(q \mid p)$, so q's marginal utility is independent of its position. Given q's independence of p, $IU(p \& q) = IU(p) + IU(q)$. Because then $IU(p) = IU(p \& q) - IU(q) = IU(p \mid q)$, it follows that $IU(p) = IU(p \mid q)$; that is, p is independent of q. Independence is symmetrical for a pair of propositions. Given that p and q are independent of each other, their marginal utilities are the same for any order of the propositions.

The norm of independence for BITs ensures that for a pair BIT_1 and BIT_2, the intrinsic utility of each BIT's object is independent of the other's object, assuming that the objects of the BITs meet the norm's conditions concerning nonproductivity. Consequently, $IU(BIT_1 \mid BIT_2) = IU(BIT_1)$ and $IU(BIT_2 \mid BIT_1) = IU(BIT_2)$, provided that the conjunction of their objects is nonproductive and so entails the object of no BIT whose object is not a conjunct. Given this condition, the norms of marginal intrinsic utility and independence entail that $IU(BIT_1 \& BIT_2) = IU(BIT_1) + IU(BIT_2)$. The independence of marginal intrinsic utilities of BITs justifies this additivity. Mutual independence of the BITs' objects ensures that the order of conjuncts does not matter.

Consider an arbitrary number of propositions. To make the intrinsic utility of their conjunction come from their marginal intrinsic utilities, order them and use the marginal intrinsic utility of each given its predecessors. To illustrate, take $(p \ \& \ q \ \& \ r)$ using the order of the conjuncts as the order of the propositions in their combination. $IU(p \ \& \ q \ \& \ r) = IU(p) + IU(q \mid p) + IU(r \mid (p \ \& \ q))$ according to a generalized norm of marginal intrinsic utility that follows from the original, binary norm by mathematical induction. The computation of $IU(p \ \& \ q \ \& \ r)$ is independent of the order of conjuncts if the marginal intrinsic utility of each conjunct is independent of any combination of other conjuncts. For example, if $IU(r) = IU(r \mid p) = IU(r \mid q) = IU(r \mid (p \ \& \ q)) = IU(r \mid (q \ \& \ p))$, then $IU(r)$ equals r's marginal intrinsic utility given its predecessors for any order of conjuncts. If the conjuncts are BITs' objects, then each conjunct's marginal intrinsic utility is independent of its predecessors for any order of conjuncts, assuming nonproductivity. So, the calculation of the conjunction's intrinsic utility may ignore the order of the conjuncts. It is just the sum of their intrinsic utilities.[26]

If a combination of propositions has more than two elements, the independence of one element from the others does not entail the independence of each from any set of the others. Addition of marginal intrinsic utilities to obtain the whole's intrinsic utility requires the more extensive network of independence relations. Granting it, the sum of marginal intrinsic utilities is the same for any order of elements and equals the sum of the elements' intrinsic utilities. The propositions' conjunction represents their combination because their order does not matter. The logical equivalence of conjunctions with conjuncts rearranged and the identity of the intrinsic utilities of logically equivalent propositions are compatible with the conjunction's intrinsic utility's being the sum of its conjuncts' marginal intrinsic utilities. Given the independence of objects of BITs, in the combination of BIT_1 and BIT_2, BIT_1's intrinsic utility plus the amount of intrinsic utility that BIT_2's realization with BIT_1 adds to BIT_1's realization equals BIT_2's intrinsic utility plus the amount of intrinsic utility that BIT_1's realization with BIT_2 adds to BIT_2's realization. Moreover, it equals the sum of the intrinsic utilities of the BITs' realizations.

An extension of intrinsic utility's additivity from pairs to n-tuples of propositions first removes marginal intrinsic utility's dependence on the

[26] For the required independence, it suffices if all but one conjunct are objects of BITs. Because of marginal intrinsic utility's additivity, if in any conjunction of propositions all, except possibly one proposition, are independent of every combination of the others, then the possible exception is also independent of every combination of the others.

order of elements in a combination by using only BITs to form combinations. Second, it moves from binary combinations to general combinations using mathematical induction. Third, it moves from nonproductive combinations to productive combinations taking advantage of entailment relations. The next paragraphs move from intrinsic utility's additivity for pairs of BITs' objects to its additivity for other combinations of BITs' objects.

The principle of additivity generalizes to the case in which the first conjunct is itself a conjunction of distinct BITs and the second conjunct is another distinct BIT. The conjunction's intrinsic utility, assuming the conjunction's nonproductivity, equals the sum of the intrinsic utility of the first conjunct and the marginal intrinsic utility of the second conjunct given the first. Given independence of the second conjunct from the first, which holds because the second conjunct is an object of a BIT, the marginal intrinsic utility of the second conjunct given the first equals the intrinsic utility of the second conjunct. Consequently, mathematical induction establishes additivity for nonproductive conjunctions of any number of objects of BITs.

According to the inductive property, a nonproductive conjunction's intrinsic utility equals the sum of the intrinsic utilities of its conjuncts, given that they are objects of BITs. The base case is a conjunction with a single conjunct. The inductive property holds trivially in this case. The inductive step moves from additivity for a conjunction to additivity for the conjunction conjoined nonproductively with a new object of a BIT, a move the norms of marginal intrinsic utility and independence support. Additivity for nonproductive conjunctions with any number of conjuncts follows from the base case and the inductive step.

Not all conjunctions of objects of BITs are nonproductive. Realization of two BITs may entail realization of a third BIT. If a conjunction of BITs' objects entails another BIT's object, then the conjunction is equivalent to its conjunction with the other BIT's object, unless the conjunction entails realization of an additional BIT. If the conjunction entails realization of an additional BIT, conjoin its object to the conjunction and continue until reaching a conjunction that entails realization of a BIT only if a conjunct entails its realization. That conjunction is equivalent to the original conjunction. Its intrinsic utility equals the sum of the intrinsic utilities of the conjuncts according to the principle of nonproductive addition, and because equivalent propositions have the same intrinsic utility, the original conjunction's intrinsic utility equals the same sum. The general principle of addition follows from the principle of nonproductive addition because productive conjunctions are equivalent to nonproductive conjunctions obtained by conjoining entailed propositions. This observation completes the theoretical argument for additivity.

A pragmatic argument, besides the theoretical argument from BITs' independence, supports intrinsic utility's additivity. Intrinsic utility directs decisions in cases in which it is equivalent to comprehensive utility and leads to a bad series of decisions if it is not additive for objects of BITs. Consider decisions in which the possible outcomes concern only BITs toward health and wisdom, and recall that a world's intrinsic utility equals its comprehensive utility. Given a failure of additivity, a person may pay more for health and wisdom together than for them separately, even though nothing makes them better or worse in combination than in isolation. The general version of this sure-loss argument for intrinsic utility's additivity considers a conjunction of distinct objects of BITs that together entail realization of no additional BIT. The conjunction's intrinsic utility should equal the sum of the intrinsic utilities of the conjuncts. Suppose that it is more than the sum. Then, the agent may purchase the conjunction for x and sell the conjuncts piecemeal for $x - \varepsilon$. The transactions produce a sure loss of ε despite being licensed by comprehensive utilities. If the conjunction's intrinsic utility is greater than the sum, a sure-loss arises from buying the conjuncts piecemeal and then selling them together. The sure-loss argument dramatizes the incoherence that arises from violating either the norm of marginal intrinsic utility or the norm of independence for BITs' intrinsic utilities.

3.6   Holism

Suppose that an agent has BITs not attached to worlds. In that case, the BITs generate the agent's intrinsic attitudes toward worlds. Holism argues against this generation. Intrinsic attitudes attaching to worlds are not any function of BITs attaching to any parts of the worlds, according to holism. Although BITs furnish reasons for intrinsic attitudes attaching to worlds, so that those intrinsic attitudes are not BITs, holism denies that BITs the worlds realize settle intrinsic attitudes toward the worlds. It rejects taking a world's intrinsic utility as the sum of the intrinsic utilities of objects of BITs the world realizes. Moreover, it rejects taking the intrinsic utility of a nonproductive conjunction of objects of BITs as the sum of the conjuncts' intrinsic utilities. This section rebuts holism's objections to intrinsic utility's additivity.

In support of welfare holism concerning lives, Raibley (2012: 241) claims, "The welfare-value of a particular occurrence within a life ought to be understood by looking at how it fits into something larger – e.g. the agent's life-plan or the agent's life as a whole." I agree that a particular occurrence's contribution to a world's intrinsic utility may similarly depend on how it fits into the world. Whether the occurrence realizes a BIT may depend on the world's other features. Nonetheless, given that in a world the occurrence is a BIT's realization, its contribution to the world's intrinsic utility does not depend on other aspects of the world. The intrinsic utilities of BITs a world realizes because of their independence add up to the world's intrinsic utility.

Holistic objections to the principle of addition generally propose counterexamples. However, typical examples do not establish that the objects of intrinsic utilities are objects of BITs and, in cases of complementarity, where realization of two BITs entails realization of a third, do not establish that additivity fails to accommodate the complementary.

A typical problem case imagines a person who likes coffee and likes tea but does not like both at once. The intrinsic utility of drinking both coffee and tea is less than the sum of the intrinsic utilities of drinking coffee and of drinking tea. This case is not a counterexample to additivity if the person's basic intrinsic desires are for the taste of tea alone and for the taste of coffee alone. These desires are not realized when drinking coffee and tea together. Furthermore, a person who has basic intrinsic desires for the taste of coffee and for the taste of tea may also have a basic intrinsic aversion to their combination. The intrinsic utility of their combination therefore sums the intrinsic utilities of realizing all three basic intrinsic attitudes. The sum may be negative. The case is not a counterexample to the addition principle because the aversion's realization explains why the beverage combination's intrinsic utility is not the sum of the intrinsic utilities of realizing the two desires. Counting all relevant considerations by counting all relevant BITs handles the beverages' complementarity. The addition principle counts all BITs the beverage-combination realizes.

May the drinker have a basic intrinsic desire for coffee in the absence of tea and a basic intrinsic desire of weaker strength for coffee in combination with tea? Because a basic intrinsic desire for coffee, including its intensity, is independent of all other BITs, it has a constant strength. An appearance of weakening arises from the productivity of its realization together with realizations of other BITs; its realization together with realization of a basic intrinsic desire for tea generates realization of a basic intrinsic aversion to the combination of coffee and tea. The apparent weakening is not genuine; the strength of a basic intrinsic desire for coffee does not change with suppositions about circumstances. The definition of BITs makes them independent so that intrinsic utilities of their realizations are constant. Alternative definitions that do not make them independent fail to generate useful addition principles.

Cases in which a basic intrinsic desire's strength increases given satisfaction of another basic intrinsic desire also challenge additivity. Suppose that an agent has basic intrinsic desires for health and for wisdom and wants wisdom more given health than given its absence. It seems that the intrinsic utility of health and wisdom does not equal the sum of the intrinsic utilities of health and of wisdom. However, in a plausible version of this case, the agent's desire for wisdom increases given health because wisdom has more benefits with health than without health. The agent has both an extrinsic and an intrinsic desire for wisdom, but only the extrinsic desire for wisdom increases given health. The agent's degree of intrinsic desire for wisdom depends exclusively on wisdom's implications. It is independent of wisdom's added benefits given health. Because of this independence, the intrinsic utility of wisdom does not increase given health, and additivity holds for the intrinsic utilities of health and of wisdom.

For another problem case, suppose that an agent has an intrinsic desire for leisure and an intrinsic aversion to having this intrinsic desire. Imagine a world that realizes both the desire and the aversion. Suppose that the world's intrinsic utility is not a sum of the intrinsic utilities of these attitudes' objects. This case does not contravene additivity unless the two intrinsic attitudes are basic and the only basic intrinsic attitudes the world realizes. However, in a rational ideal agent that the principle governs, the two attitudes are not both basic. They are not independent because one attitude is a reason for a feature of the other. The intrinsic aversion to desiring leisure diminishes the intrinsic desire for leisure. Cases in which complementarity seems to generate counterexamples to additivity either incorrectly identify BITs or else ignore some BITs.[27]

Does additivity have a get-out-of-jail-free card? Any reason against a conjunction's additivity just puts another intrinsic utility in the sum of intrinsic utilities that, according to additivity, equals the conjunction's intrinsic utility. The sum's adjustability may make additivity a trivial truth. That is not an objection to its truth, however. Consider an application of the rule of inference *modus ponens*. If the conclusion is not true, then either the conditional is false or its antecedent is false. That sure-fire defense of *modus ponens*'s validity does not defeat the rule of inference. Having a response to any objection does not count against the rule. Additivity is a

---

[27] I thank Troy Nunley (personal communication, 2001) for insightful points about cases involving higher-order intrinsic attitudes.

normative principle for an ideal agent's degrees of intrinsic desire. The normative principle has the support of judgments that an ideal agent's violations of the principle, although possible, are irrational. The normative principle if true is necessarily true and resistant to all putative counter-examples. It uses an agent's BITs to ward off counterexamples and does not fabricate BITs an agent lacks just to block counterexamples.

Intrinsic utility shares some structural features with (objective) intrinsic value as moral theory interprets it.²⁸ Some value theorists recognize organic unities that have an intrinsic value not equal to the sum of the intrinsic values of their parts.²⁹ For example, suppose in some situation a person is suffering. An addition to that situation is another person's taking pleasure in the first's suffering. The addition makes the situation worse. So, the value of the second situation is not a sum of the value of its parts, the objection contends. Adding something of value, pleasure, to the original situation lowers its value. This lowering is incompatible with value's additivity. If the intrinsic value of the whole were a sum of the intrinsic values of the parts, adding pleasure, a good, would make the original situation better, not worse.³⁰

Several points distinguish intrinsic utility from intrinsic value. First, intrinsic utility rests on psychological states, not values, and has a structure imposed by rationality, not morality. Second, relations among intrinsic attitudes rather than relations among the attitudes' objects characterize BITs that form the foundation of intrinsic utility. Intrinsic values do not have the causal structure that intrinsic attitudes have. Desires cause other desires, but values do not cause other values. Third, the debate about objects of intrinsic value – states of affairs, propositions, or concrete objects – does not arise for the objects of intrinsic desires and aversions.³¹ Desire and aversion are propositional attitudes and have propositions as objects. The objects of intrinsic utilities are propositions and so are finely individuated. A proposition expressing experience of a bad pleasure is not

---

²⁸ Basic intrinsic attitudes have a foundational role similar to the foundational role of basic intrinsic values in Feldman's (2000) and Zimmerman's (2001: chap. 5) theories of intrinsic value.

²⁹ Moore ([1903] 1993: 78–80) proposes a principle of organic unities that states that the intrinsic value of a whole need not equal the sum of the intrinsic values of its parts. Chisholm (1986: chap. 7) and Dancy (2004: chap. 10) treat assessment of organic unities.

³⁰ Dancy (1993: 61) states objections to value's additivity along the lines of the case of the bad pleasure. Chisholm (1986) notes Brentano's view that sadomasochists are averse to pain but attracted to something that has pain as a part. Chisholm takes pleasure to be part of bad pleasure. Additivity does not apply to a person's bad pleasure using pleasure as a part. In contrast, Zimmerman (2001: 148) denies that pleasure is part of bad pleasure.

³¹ Lemos (1994) and Rabinowicz and Rønnow-Rasmussen (2000) review this debate.

the same object as a proposition expressing experience of a pleasure, even if the bad pleasure realizes the pleasure. Fourth, an analysis of intrinsic utility, unlike an analysis of intrinsic value, may apply only to agents meeting certain assumptions, in particular, assumptions about their rationality and cognitive abilities. Because of these differences, even if intrinsic values are organic unities, intrinsic utilities may obey the addition principle.[32]

Additivity for a world's intrinsic utility recognizes cases in which a BIT attaches to a world, so that the world is an organic unity. Additivity depends on BITs. It does not deny organic unities. It just advances a principle of addition that analyzes a world's intrinsic utility when BITs attach to its proper parts. The example of the bad pleasure does not refute intrinsic utility's additivity. Intrinsic utility attaches to a proposition (given a way of understanding it). The example adds to a person's suffering another person's taking pleasure in the first's suffering, not just pleasure. For a typical person, pleasure in another's suffering is intrinsically unattractive, not attractive. The addition of the unattractive pleasure lowers the situation's intrinsic utility. The case of the bad pleasure is not a counterexample to intrinsic utility's additivity.

Another objection to additivity uses an example Parfit (1986: 161) constructs. He imagines a century of ecstasy and a drab eternity. The century of ecstasy has more intrinsic utility than the drab eternity even though every day of the drab eternity has the same small, positive intrinsic utility. If intrinsic utilities were additive, the drab eternity would have more intrinsic utility than the century of ecstasy. Hence, intrinsic utility is not additive, the objection claims.

A judgment that the century of ecstasy has more intrinsic utility than the drab eternity assumes that the drab eternity is boring, whereas the century of ecstasy is exciting. This suggests a reply to the objection. A series of spread out, small, exactly similar pleasures realizes a basic intrinsic aversion, whereas a series of closely packed, intense, diverse pleasures realizes a basic intrinsic desire. The BITs the two series realize explain the comparison of their intrinsic utilities.

The final objection to additivity is an abstract case. If an agent's BITs overdetermine his intrinsic attitudes, then the intrinsic utility of a

---

[32] Suppose that a fully informed agent cares only about intrinsic value. For this agent, a proposition's intrinsic utility coincides with its intrinsic value. If intrinsic value is holist, then intrinsic utility for this agent is also holistic. However, not every ideal agent, not even every rational ideal agent, cares only about intrinsic value. An ideal agent's cognitive abilities do not ensure virtue, and rationality does not ensure morality. For some ideal agents, intrinsic utility and value diverge. Intrinsic utility need not be holistic for such agents even if value is holistic.

conjunction of BITs' objects need not equal the sum of the intrinsic utilities of the BITs whose realization it entails. The sum may double count considerations. Suppose, for example, that no intrinsic desires ground the intrinsic desires that p, that r, that p & q, and that q & r, so that they are all basic. $IU(p$ & q & $r) \neq IU(p) + IU(p$ & $q) + IU(q$ & $r) + IU(r)$, contrary to additivity. The sum double counts p, q, and r. This objection fails because the BITs the case supposes are not all genuine. Because of entailments among their objects, the intrinsic desires that p and that p & q are not both basic. The desire that p supplies a reason for the desire that p & q, so that the second desire is not basic. Similarly, the intrinsic desires that r and that q & r are not both basic. At least one of these intrinsic desires derives from other intrinsic desires. If the intrinsic desire that r is basic, then the intrinsic desire that q & r derives from it.

3.7 Separability

Let us take a world part, such as the future, as a variable to be given a value. A proposition represents its value. The *complement* of a world part is another world part such that a value for the first world part and a compatible value for the second world part combine to form a whole world. The proposition representing a world part's value conjoined with the proposition representing the value of the world part's complement represents the world. A world part's value is distinct from a proposition that represents it, although for brevity I sometimes call a world part's value a proposition.

Common methods of simplifying choices assume the separability of world parts such as the past and the future. This section shows that some divisions of a world yield parts that are separable from their complements when intrinsic utility settles order. The separability of a part from its complement is equivalent to the intrinsic utility of a world's being a function of the part's intrinsic utility and its complement, with the function being strictly monotonically increasing with respect to the part's intrinsic utility, so that the order of worlds is the same as the order of the part's values given any way of fixing its complement.

This section's claims about separability apply within a model that incorporates idealizations. Its argument for separability presents reasons why an ideal agent with BITs toward values of world parts should have a separable intrinsic-utility function, according to which intrinsic utilities of worlds separate into intrinsic utilities of values of world parts. The argument uses the independence of BITs to establish a norm of separability for ideal agents. The model displays the operation of some factors that explain rational choice and thereby improves utility theory's position for discovering general principles free of idealizations.

Support for the norm of separability comes in two forms. First, separability follows from compliance with the norm of additivity. Second, separability follows directly from compliance with BITs' norm of independence, without assuming that their independence supports additivity. The second argument supplements Section 3.5's support for the norm of additivity because without appeal to additivity, it establishes as a norm separability, a consequence of additivity.[33]

Intrinsic utility's additivity assists calculations of utilities but is not necessary for justifying simplifications of choice. Section 3.5 argues for the traditional principle of additivity because of its independent interest, and not just because it establishes a world part's separability from its complement. The direct argument from BITs' independence to separability, without passing through additivity, accommodates agents with intrinsic preferences but without quantitative intrinsic desires and aversions. It anticipates generalization of this chapter's normative model.

Chapters 4–6 need only separability of selected world parts from their complements. If the world part evaluated is x and its complement is $-x$, then simplification of choice requires showing that $IU(-x, x) = F(-x, IU(x))$ and that F is strictly monotonically increasing with respect to $IU(x)$, or equivalently that x is separable from $-x$. It does not need mutual separability, or even compositionality. Moreover, it need not establish a world part's separability in general but only its separability for a particular way of fixing the world part's complement, in the case of the future's separability, fixing the variable for the past at its actual value (although establishing the future's separability in general permits simplification of choice when the past's actual value is unknown).

For general coverage of the separability that Chapters 4–6 assume, this section nonetheless shows that an arbitrary world part is separable from its complement. Showing this establishes as a by-product *weak separability*, that is, the separability of each world part from its complement. Moreover, it establishes *strong separability*, that is, the separability of each set of world parts from its complement, because a set of world parts forms a world part.

---

[33] Apologizing for bad behavior is a norm, and compliance entails bad behavior, although no norm requires bad behavior. Showing that compliance with a norm entails meeting some constraint does not by itself establish that meeting the constraint is a norm. The demonstration needs supplementation to support the norm conclusively.

Separability holds with respect to variables, their values, and an order of their values, or a utility function representing their order. Separability of intrinsic utility depends on specification of the relevant variables and their values. The variables settle structure; they identify a composite's parts. Familiar forms of separation use variables that represent world parts such as the past and the future. A demonstration of separability for an arbitrary world part selects an arbitrary division of a world into suitable exclusive and exhaustive parts, selects an arbitrary part from the division, and shows that the part is separable from its complement.

The world parts this section treats are variables having as values realizations of sets of BITs. A conjunction of the objects of BITs in a set represents the set. The smallest world part is an indicator variable for a BIT with values that indicate whether the BIT is realized. Its values are either the unit set with the BIT's realization or the empty set. A world part's values may come from an adjustable number of indicator variables for BITs. For example, a part may have as its value the realization of BIT_1, or the realizations of BIT_1 and BIT_2. A world part's complement is a variable whose value together with the world part's value composes a world. Hence, the past's complement is the future (taken to include the present). Table 3.1 shows a vector of variables' values that represents a world realizing three BITs.

This section treats as variables an arbitrary world part and its complement, represents the values of these variables with conjunctions of objects of BITs, and uses intrinsic preference to order the values of these variables. It establishes this principle of separability: *an arbitrary world part is separable from its complement, and a world's intrinsic utility is consequently a separable function of the intrinsic utilities of the world part and its complement.*

The separability of intrinsic utilities follows from the separability of world parts because in the ideal agents this section treats, intrinsic utilities agree with intrinsic preferences. Later chapters use the italicized principle to support for options' worlds the separability of the future from the past, the separability of the options' region of influence from its complement, and the separability of options' consequences from their complement. Because

Table 3.1 *A world and its parts*

Variables	Past	Future
values	$\{BIT_1, BIT_2\}$	$\{BIT_3\}$
vector	$(\{BIT_1, BIT_2\},$	$\{BIT_3\})$

they define temporal, spatiotemporal, and causal utility using intrinsic utility, they derive the separability of these types of utility from the separability of the world parts from their complements.

Because the separability of intrinsic utilities of realizations of BITs is fundamental rather than derived from another quantity's separability, an argument for it rests on basic principles and generalization from cases. The first argument moves from additivity of values of variables to additivity of variables, and then to separability of variables. Section 3.5 establishes additivity of intrinsic utilities of realizations of BITs. This section uses their additivity to establish intrinsic utility's additivity for variables with realizations of BITs as values and then establishes the variables' separability.

Probability's additivity for disjunctions is a generalization involving variables for disjuncts that entails additivity for an exclusive disjunction that values of the variables compose. Intrinsic utility's additivity for a world involves variables for conjuncts that together form a world's representation. That they form a world's representation requires that their values be comprehensive and consistent. Additivity of variables is additivity of the values of the variables, one value for each variable, for every instantiation of the variables that makes a world.

For world parts, additivity of values does not logically entail additivity of variables because conjunction may be productive and disrupt additivity of variables. It is logically possible that the intrinsic utility of a BIT's realization changes from world to world so that although a world's intrinsic utility is additive, interchangeability, a consequence of additivity of world parts, does not hold. Two values of a part that have the same intrinsic utilities may interact differently with values of other parts. This may block the parts' additivity. Additivity for a conjunction of objects of BITs by itself does not yield additivity of the variables representing world parts. World parts, the variables, may have as values different subsets of BITs' realizations. Replacing a value with one of equal intrinsic utility may change world intrinsic utility by changing the number of BITs realized.

Making the values of variables comprehensive makes additivity of values yield additivity of variables. Additivity of values holds for variables' comprehensive values. A variable's value may depend on other variables' values. A variable may not take all its possible values in all possible contexts. A variable's comprehensive value accommodates the effect on its value that the values of other variables exert. A variable's comprehensive value may change given changes in the values of other variables. Additivity of variables does not require that the variables' values be independent of other variables' values.

A variable's comprehensive value includes what it produces when conjoined with other variables' values to create a world. Given an interchange of a variable's value holding other variables fixed, all BITs the interchange realizes belong to the value of the variable changed. Given that the original and the new value have the same intrinsic utility, the world's intrinsic utility is constant. Let the parts of a binary conjunction representing a world be the first conjunct and the second conjunct, each itself a conjunction of BITs' objects. Suppose that replacing a value of the second conjunct with a new value is productive. Any realization of a BIT that the conjunction but not the first conjunct entails becomes part of the second conjunct.

Intrinsic utility's additivity for a conjunction of objects of BITs uses intrinsic utilities of objects of all BITs whose realization the conjunction entails. It does not use only the conjuncts' intrinsic utilities. Swapping objects of BITs with equal intrinsic utilities may change the conjunction's intrinsic utility because the new object of a BIT combines with the others to entail realization of another BIT. Additivity of objects still holds for the new conjunction. A BIT's object makes an independent contribution to the conjunction's intrinsic utility even though the BIT's realization may interact with the other BITs' realizations to generate realization of additional BITs. Its independent contribution grounds intrinsic utility's additivity despite the productivity of its realization. The additivity of objects of BITs grounds the additivity of world parts that have sets of objects of BITs as values, given the values' comprehensiveness.

Suppose that an agent cares only about pleasure and relations among pleasures. If an agent has a basic intrinsic desire for continuity of pleasure, then a second pleasure following and equaling a first pleasure not only increases the world's intrinsic utility by adding the second pleasure but also increases the world's intrinsic utility by introducing continuity of pleasure. For a world, every BIT it realizes falls into just one part of the world. If changing the value of a world part to add a BIT's realization entails another BIT's realization, then the comprehensive value of the world part has not just the realization of the BIT added but also the realization of every other BIT that the addition entails in the context of the value of the world part's complement. If the value of a world part includes the first pleasure and the value of the world part's complement includes the second pleasure, then adding the second pleasure to the value of the complement also adds continuity of pleasure to the value of the complement. In general, if a world part and its complement are productive, the product belongs to the part that is second in the order of world parts.

Consider the past and future, taken as variables. Replace the future's value with another holding the past fixed. Suppose that the new future, unlike the old, continues a tradition, which realizes a basic intrinsic desire. The change is productive. Realizing the desire is part of the world and part of its future taken comprehensively. Additivity of values of variables holds for the original world and for the new world. Because of this additivity, additivity of variables for the past and the future also holds. $IU(w)$ = $IU(\text{past}) + IU(\text{future})$, where past and future are variables. A value of the variable for the future is comprehensive and includes objects of BITs that a future realizes given the past's value. If a past and future entail a BIT's realization that the past does not entail, then the future taken comprehensively entails it.

Additivity of values yields additivity of variables because the values of the variables are comprehensive. Given the additivity of the intrinsic utilities of world parts taken as variables, separability of a world part from its complement follows. The order of worlds according to intrinsic utility is the same as the order of the part's values according to intrinsic utility, given any way of fixing its complement's value.

Suppose that the future is separable from the past because a world's intrinsic utility equals the intrinsic utility of the world's future divided by its past's intrinsic utility: $IU(w) = IU(\text{future})/IU(\text{past})$. Holding the past fixed, the order of futures settles the order of worlds. In this case, the past is not separable from the future, however. Holding fixed the future's intrinsic utility, increasing the past's intrinsic utility lowers the world's intrinsic utility. Suppose that $IU(w_1) = IU(\text{future}_1)/IU(\text{past}_1) = 4/2$, and $IU(w_2) = IU(\text{future}_2)/IU(\text{past}_2) = 4/4$. Then, keeping the future's intrinsic utility fixed at 4 and increasing the past's intrinsic utility from 2 to 4 decreases the world's intrinsic utility from 2 to 1. For a fixed future, the order of pasts is contrary to the order of worlds. A variable's separability requires that the order of its values be the same given any way of fixing the values of the other variables, and, as noted, weak separability requires this independence for each world part and its complement. Establishing the weak separability of world parts requires more than establishing the separability of one world part from its complement. Weak separability nonetheless follows from the demonstration that an arbitrary world part is separable from its complement.

The demonstration of separability for an arbitrary world part also establishes strong separability of world parts: every set of world parts is separable from its complement. Strong separability follows because a set of world parts forms a world part. From the additivity of exclusive and exhaustive world parts x, y, z, it follows that $IU(x, y, z) = IU(x) + IU(y) + IU(z)$. It also follows that $IU(x, y, z) = IU(x, y) + IU(z)$ because $IU(x, y) = IU(x) + IU(y)$. As $IU(x, y)$ increases, $IU(x, y, z)$ increases. This establishes the separability of $IU(x, y)$ from z in a function going from them to $IU(x, y, z)$, and equivalently the separability of $\{x, y\}$ from $\{z\}$. Increasing the set of parts' intrinsic utility increases the world's intrinsic utility holding other parts fixed. Worlds' intrinsic utilities are strictly monotonically increasing with respect to their parts' intrinsic utilities.

Given the additivity of exclusive and exhaustive world parts x and y, $IU(x, y) = F(IU(x), IU(y))$ where F is addition. In $F(IU(x), IU(y))$, the utilities $IU(x)$ and $IU(y)$ are mutually separable, or weakly separable, and because their combination has just two variables, they are also strongly separable. Because the variables are intrinsic utilities, they are functionally separable. Mutual separability of $IU(x)$ and $IU(y)$ is equivalent to mutual separability of x and y when the order of intrinsic utilities gives the order of values of variables. In the rational ideal agents this chapter treats, the order of intrinsic utilities agrees with the order of intrinsic preference (even if intrinsic utility is not defined by intrinsic preference).[34]

A second argument for separability, a comparative norm that additivity entails, relies directly on BITs' independence rather than on additivity. It also shows that an arbitrary world part is separable from its complement. That is, for the world part, the ranking of the world part's values is independent of its complement's value.

A variable in a vector of variables is separable from its complement if and only if the ranking of its values is the same for any assignment of values to the variables in its complement. An agent's ranking of a variable's values given an assignment of values to other variables assumes that the agent knows the assignment and compares values' contributions given the assignment. The ranking compares values' marginal contributions, that is, their contributions to composites that begin with the assignment. Given that a

---

[34] Additivity of variables and the separability it entails yields compositionality. Suppose that an agent has n BITs and that x_i is an indicator variable specifying whether a world w realizes BIT_i. Additivity for variables holds if and only if $IU(w) = \sum_i IU(x_i)$. (Section 3.5's formula ignores BITs that w does not realize because their intrinsic utilities equal zero.) Compositionality of a world's intrinsic utility obtains if and only if for some function F, $\cdot IU(w) = F(IU(x_1), IU(x_2), \ldots, IU(x_n))$. An equivalent property of interchange asserts that if w' is like w except for realizing BIT_j instead of BIT_i, BIT_j is distinct from all other BITs that w' realizes, and $IU(BIT_j) = IU(BIT_i)$, then $IU(w) = IU(w')$. This type of interchange is nonproductive. Replacement of realization of BIT_i with realization of BIT_j does not interact with other features of w' affecting its intrinsic utility. If it did, then w' would not be like w except for swapping one BIT's realization for another's realization. Replacement would affect an additional BIT's realization. The restrictions duplicate the effect of limiting interchange to comprehensive values of variables.

first glove is a right glove, a second glove's marginal contribution typically is greater if it is a left, rather than a right, glove. The comparison flips if the first glove is a left glove. In contrast, given that lunch is a sandwich, a monetary gain's marginal contribution typically is greater if it is $5 rather than $1, and the comparison of monetary gains is the same if lunch is soup. A variable's separability from other variables entails a constant ranking of its values and so is a feature that the variable for the second glove lacks and that the variable for monetary gains possesses.

The argument for a world part's separability assumes a world's division into exclusive and exhaustive world parts, with world parts being variables that have sets of BITs' objects as values. Setting the values of the parts makes a complete world, and no BIT's object falls into two values contributing to a single world. An agent's decision problem settles the possible values of world parts, such as the past and the future. The future's values are the sets of BITs whose realizations the future may bring according to the option the agent adopts.

Suppose that a tourist has a basic intrinsic desire to visit Venice at least once and considers making a trip to Venice. Satisfying the desire is part of a possible future, assuming that the tourist has not already visited Venice. Fixing the past so that the tourist has already visited Venice removes the desire's satisfaction from the possible futures. The value of the past may change the possible values of the future. This happens if given one way of fixing the past, a possible value of the future is $\{BIT_1\}$, and given another way of fixing the past that includes realization of BIT_1, $\{BIT_1\}$ is no longer a possible value of the future. Setting the value of the past may affect the possible values of the future because the values of world parts combine to form worlds. A world part is separable from other world parts if and only if for all ways of fixing the values of the other world parts, its compatible values have the same ranking.

The values of a world part are comprehensive. Given an assignment of values to the part's complement, the part's value includes every BIT's realization, in the world its value and its complement's value constitute, that its complement's value does not include.

The argument for a world part's separability uses a general principle stating that intrinsic preferences among realizations of sets of BITs are independent of circumstances of evaluation. The case for this principle, following Section 3.3's case for a similar principle, observes that the reasons for a BIT depend only on its realization's implications. Hence, the intrinsic utility of a set of BITs' realization depends only on its realization's implications. These implications are independent of circumstances. The independence of BITs entails the independence of intrinsic preferences among realizations of sets of BITs. An intrinsic preference between marginal contributions of realizations of two sets of BITs is independent of circumstances. The preference is the same given any condition. The only provisos are, first, that the condition not entail the object of any BIT in either set and, second, that the realization of the condition and each set of BITs be nonproductive. If the condition is realization of another set of BITs making a world part's value, then the nonentailment proviso is met. If the other set of BITs' realization entails the object of any BIT, the BIT's realization belongs to the other set because a world part's values are comprehensive. Also, the nonproductivity proviso is met because the values of a world part are comprehensive. Given an assignment of values to the part's complement, the part's value includes every BIT's realization, in the world its value and its complement's value constitute, that its complement's value does not include.

A world part's values are sets of BITs. Its complement's value settles the BITs a world realizes except for the BITs that the world part realizes. Given the general principle, for any value of its complement, the sets of BITs that are the world part's values have the same ranking. No change in its complement's value changes the ranking because the reasons for the ranking are independent of its complement's value. Suppose that given a way of fixing the BITs that the past realizes, the ranking of sets of BITs that the future might realize is $\{BIT_1\} > \{BIT_2, BIT_3\} > \{BIT_4\}$. Then, given another way of fixing the BITs that the past realizes, a way that does not change the sets of BITs that the future might realize, their ranking remains the same. If the other way of fixing the past changes the sets of BITs that the future might realize, then any sets that remain possible have the same ranking.

The last step in establishing separability is showing that the constant ranking of a world part's values agrees with the ranking of worlds that contain the part's values and share the same value for the part's complement. This agreement holds because comparisons of marginal contributions agree with comparisons of worlds. A rational ideal agent intrinsically prefers one marginal contribution to another if and only if she prefers a world with the first contribution to a world with the second contribution if the two worlds include the same value for the part's complement. Hence, the order of the part's values agrees with the order of worlds.

Suppose that realizations of sets of BITs have intrinsic utilities. In this quantitative environment, with world parts having realizations of sets of BITs as values, a world part x is separable from its complement $-x$, if given any value of $-x$, for some positive α, $IU(x \mid -x) = \alpha IU(x \,\&\, -x)$ for all values of x. This sufficient condition for separability states that the marginal intrinsic utility of a part's value given a fixed value for its complement is proportional to the intrinsic utility of the world the part's value creates when joined with the complement's value. Also, x's separability from $-x$ is equivalent to strictly increasing monotonicity in the second argument place for a function $F(-x, IU(x))$ yielding a world's intrinsic utility. If the part x's intrinsic utility increases because the part realizes an additional basic intrinsic desire, then the world realizes an additional basic intrinsic desire, and its intrinsic utility increases. Similarly, if the part x's intrinsic utility decreases because the part realizes an additional basic intrinsic aversion, then the world realizes an additional basic intrinsic aversion, and its intrinsic utility decreases.

The independence of the part x's intrinsic utility follows from the independence of its values' intrinsic utilities. Each value's intrinsic utility is constant for all values of the other variables. Because the part x's value and other variables' values are not complementary, the intrinsic utility of the variable's value is independent of the other variables' values. Independence holds because the intrinsic utility of a set of BITs' realization is the same in all contexts. It is the set's constant contribution to the intrinsic utility of worlds that realize it.

For each world part x, $IU(x)$ is independent of $-x$. So the order of $IU(x)$'s values is independent of $-x$. In $F(-x, IU(x))$, the variable $IU(x)$ is separable from $-x$ because the marginal contributions of x's values have the same ranking as the worlds that contain them. This separability of intrinsic utility holds because in a world of the form $(-x, x)$, the variable x is separable from the variable $-x$, and intrinsic utility agrees with the orders of the vectors and subvectors of variables' values.

Separability is a type of independence. So, using independence to argue for separability may appear circular, especially because in the quantitative case, marginal intrinsic utility may characterize both independence and separability. However, the independence of reasons for a BIT supports, as a norm, the independence of the BIT's marginal intrinsic utility. The difference in types of independence prevents circularity.

3.8   Incomplete information

Using intrinsic-utility analysis to calculate the intrinsic and comprehensive utilities of an option's world requires identifying the BITs the world realizes and so suits cases with certainty of an option's world. When an option's world is uncertain, its comprehensive utility is a probability-weighted average of the comprehensive utilities of the worlds that might be its world. The comprehensive utilities of these worlds equal their intrinsic utilities. Expected-utility analysis, in the form of world-Bayesianism, combines with intrinsic-utility analysis.

The various forms of utility analysis are consistent because they derive from a canonical form of utility analysis and canonical ways of obtaining inputs for the forms of utility analysis. The derivations assume, as do the forms of utility analysis, that agents are rational and ideal as well as in ideal decision problems. A rational ideal agent is cognitively perfect, fully meets rationality's requirements, and in an ideal decision problem assigns probabilities and utilities sufficient for conducting utility analyses and maximizing utility. A restriction makes finite the number of BITs and so the number of possible worlds (trimmed of irrelevancies).

The canonical method of computing utility is a combination of expected- and intrinsic-utility analyses. It combines expected-utility analysis using possible worlds and intrinsic-utility analysis of possible worlds using BITs they realize. $U(o) = \sum_i P(w_i \text{ given } o) \sum_{j \in \{ij\}} IU(BIT_j)$, where $\{i\}$ indexes possible worlds and $\{ij\}$ indexes the objects of the BITs w_i realizes. In other words, the method takes the comprehensive utility of an option as a probability-weighted average of the sums of the intrinsic utilities of the sets of BITs that might be realized if the option were realized. All methods of calculating an option's comprehensive utility agree with this method and so are consistent with one another. The consistency results extend to all propositions, but this book's demonstrations focus on options because its primary topic is utility analysis's application in decision problems.

The canonical methods of obtaining input for utility analyses draw on probabilities of worlds and intrinsic utilities of objects of BITs. For instance, they derive a state's probability in the usual way by summing the probabilities of the worlds in the set of worlds that represents the state.

Weirich (2001: appendix) shows that expected- and intrinsic-utility analyses derive from the canonical form of utility analysis. Two features of worlds' utilities unify the addition principle and the expected-utility principle for options. First, a world's comprehensive utility equals its intrinsic utility because a world fully specifies every significant event it realizes. Second, the worlds' comprehensive utilities generate an option's expected utility using possible worlds to form a partition of possible outcomes.

The appendices to the chapters presenting temporal-, spatiotemporal-, and causal-utility analyses verify their applications' consistency with the canonical form of utility analysis. Because intrinsic utilities define temporal, spatiotemporal, and causal utilities, the demonstrations of consistency reduce these types of utility to intrinsic utility without reliance on rules of rationality for the ideal case, in contrast with the demonstrations of consistency for methods of obtaining comprehensive utilities, which intrinsic utilities ground but do not define.

3.9   Measurement

Degrees of intrinsic desire may exist even if they are not measurable, but their measurement facilitates applications of intrinsic-utility analysis. Although their measurement may use either the causes or effects of intrinsic attitudes, most methods rely on their effects. In an ordinary agent, degrees of intrinsic desire may be unstructured. Lack of structure complicates their measurement by effects. Psychometrics provides methods of measuring intrinsic attitudes using reports that intensities of intrinsic desires and aversions prompt.

An agent may intrinsically desire one proposition's realization twice as much as another proposition's realization. For example, a miser obsessed with money may intrinsically want $2 twice as much as $1. In a typical case, reflection reports approximately the intensity of an intrinsic desire. Reflection's report that one intensity is twice another offers nonconclusive evidence that the intensities are as reported. Degrees of intrinsic desire are numerical representations of quantitative empirical relations that reflection may assess. A convenient measure of an agent's intrinsic desire is the central tendency of her reports about how many times more intense the intrinsic desire is than another intrinsic desire that serves as a unit. In some cases, reflection is unreliable, but these cases do not discredit it generally. If agents are ideal, as this section assumes, they have reliable reflective access to their attitudes.[35]

For a rational ideal agent, measurement of intrinsic utility may use the norms that govern it, such as the additivity of a world's intrinsic utility. Intrinsic utilities, a rational ideal agent's degrees of intrinsic desire, have a structure that makes them measurable in some cases. Methods of measuring

---

[35] A useful measure of a quantity need not be either 100 percent reliable or 100 percent accurate. As a bathroom scale does for weight, it may provide inconclusive evidence of the quantity to be measured within a margin of error. Loewenstein and Schkade (1999) describe people's tendencies to make mistakes predicting their feelings. Robinson and Clore (2002) describe discrepancies between people's reports of their current emotions and their predictions and recollections of their emotions. However, these theorists acknowledge that in most cases, people correctly report past and future emotions and feelings. Their studies do not discredit the general reliability of people's reports about their current desires.

a rational ideal agent's intrinsic desires may have satisfactory accuracy for real agents who over a limited range of intrinsic desires approximate rational ideal agents.

Worlds' intrinsic utilities equal their comprehensive utilities, and the latter are derivable from preferences among gambles, as in Savage ([1954] 1972: chap. 3).[36] The method of derivation may use preferences among hypothetical gambles that actual choices do not manifest. The intrinsic utilities of worlds, given structural assumptions, imply the intrinsic utilities of objects of BITs. Hence, preferences among gambles ground inferences of BITs' intrinsic utilities.

In experimental situations in which an agent may choose between chances for realizations of basic intrinsic attitudes, expected-utility analysis applies. Suppose that an agent may realize a basic intrinsic desire for a pleasure and no other intrinsic desire. She may also gamble in a way that yields exactly realization of a basic intrinsic desire for a second pleasure if heads turns up after a toss of a symmetric coin and otherwise yields realization of no basic intrinsic attitude. She is indifferent between the two options. Suppose that realization of the second pleasure serves as a unit for intrinsic desire and that realization of no basic intrinsic attitude serves as a zero point. Then, the intrinsic desire for the first pleasure is 0.5 on this scale.

To illustrate additional types of inference, grant that health and wisdom are objects of basic intrinsic desires and hence BITs. Suppose, moreover, that only these BITs exist so that trimmed worlds specify just whether these BITs are realized. Grant that intrinsic utilities of worlds are inferred from comprehensive utilities of worlds measured using preferences among options, following Savage's representation theorem. Imagine that $IU(H \& W) = 2$, $IU(H \& {\sim}W) = IU({\sim}H \& W) = 1$, and $IU({\sim}H \& {\sim}W) = 0$. Then, $IU(H) = IU(W) = 1$, assuming additivity. Provided that a BIT's nonrealization has zero intrinsic utility, that $IU(H \& {\sim}W) = 1$ suffices for an inference that $IU(H) = 1$. Similarly, that $IU({\sim}H \& W) = 1$ suffices for an inference that $IU(W) = 1$. Alternatively, from $IU(H \& W) = 2$ and $IU(H) = 1$, it follows that $IU(W) = 1$, given additivity.

---

[36] Binmore (2009) and Gilboa (2009: chap. 10) review Savage's representation theorem. A utility is state independent if and only if its utility given a state is the same for all states. The absence of state-independent, comprehensive utilities for propositions less specific than a possible world impedes comprehensive utility's measurement, as Schervish, Seidenfeld, and Kadane (1990) show. Intrinsic utilities of propositions are state independent, but their state independence does not entail the state independence of the propositions' comprehensive utilities.

A representation theorem for preferences generalizes this pattern of inference. It shows the existence and uniqueness of a representation obeying the principle of additivity and has the same form as a representation theorem for quantities such as weight and length. The representation theorem has structural assumptions that limit its application, as do other representation theorems.[37]

A representation of preferences among options using intrinsic utility shows that the preferences agree with an assignment of intrinsic utilities to worlds generated from intrinsic utilities assigned to realizations of BITs selected to generate by addition the worlds' intrinsic utilities. The representation shows that the preferences are "as if" maximizing comprehensive utility, granting that realizations of BITs generate worlds' intrinsic utilities and so options' comprehensive utilities. A representation theorem shows that under certain assumptions, a representation of preferences using intrinsic utility exists and is unique except for scale changes. The representation does not define intrinsic utility, which is degree of intrinsic desire.

A representation theorem shows only that given its assumptions, comparisons of worlds are "as if" the worlds' intrinsic utilities come from an additive combination of their components' intrinsic utilities. The results are stated hypothetically but are real if worlds' intrinsic utilities derive from their components' intrinsic utilities. In some cases, only an additive combination of parts' intrinsic utilities yields worlds' intrinsic utilities. Then, the parts' intrinsic utilities really generate, and not just hypothetically generate, the worlds' intrinsic utilities. The norm of strong separability for an ideal agent grounds inference of intrinsic utilities of composites and their components from the agent's preferences, including conditional preferences. Compliance with the norm ensures the existence and uniqueness of an additive representation (as Section 1.3 notes), and this representation, given additivity, yields genuine intrinsic utilities.

Measurement of intrinsic utilities using worlds' comprehensive utilities is derivative and not fundamental measurement. As Section 1.2.2 explains, adopting a concatenation operation for intrinsic utilities makes possible their fundamental measurement. Fundamental measurement of length adopts a concatenation operation that addition of lengths represents. For straight sticks, end-to-end juxtaposition in a straight line may be length's concatenation operation. Fundamental measurement of weight similarly adopts a concatenation operation. For stones, putting them together in a

---

[37] Krantz et al. (1971: chap. 6) and Baron (2008: chap. 14) describe the methods of conjoint measurement that yield such representation theorems.

balance pan may be weight's concatenation operation. A concatenation operation's adoption stipulates additivity with respect to the operation. It grounds measurement of extensive properties such as length and weight that vary with division of objects possessing them, as opposed to intensive properties such as temperature that do not vary with division of objects possessing them. The representation theorems for the fundamental measurement of length and weight in Krantz et al. (1971: chap. 3) apply to intrinsic utility given a concatenation operation for it.

The concatenation operation for intrinsic utility is nonproductive conjunction of BITs' objects. It makes intrinsic utilities measurable and "defines" them operationally if desired. Measurement of the intrinsic utility of a BIT's object may use nonproductive conjunction of the objects of two other BITs of equal intensity whose conjunction has the same intrinsic utility as the first BIT's object. This method assumes discoverability that two unknown intrinsic utilities are equal. Suppose that IU(the pleasure of a sip of water) + IU(the pleasure of stretching legs) = IU(the pleasure of a sip of water and the pleasure of stretching legs) = IU(knowing Oregon's capital). If the intrinsic utilities summed are equal, then in the string the last intrinsic utility is double the first. The concatenation operation generates quantitative comparisons of intrinsic utilities.

A collection of units to concatenate assists measurement by comparison with concatenations. Measurement of intrinsic utility cannot use for comparisons concatenations of the same proposition because (p & p) is equivalent to p. Concatenation of the same proposition does not increase intrinsic utility. For intrinsic utility, units are objects of distinct but equivalent BITs that increase intrinsic utility when conjoined to other such objects or their conjunctions. Measurement may conjoin them to form "measuring sticks," assuming that conjunction is nonproductive.

Measurement may affect objects to be measured. Conjoining the objects of two BITs may entail a third's object. Strong structural assumptions may remove such complications. This section's technique for fundamental measurement restricts the concatenation operation to BITs whose realizations are unproductive in concatenations. A nonproductive conjunction of BITs' objects entails only objects of BITs that some conjunct entails. However, inferring intrinsic utilities of BITs' realizations is possible despite productivity because a BIT has a constant intensity in all realizations. Future research may relax restrictions and use mathematical ingenuity to generalize techniques of fundamental measurement.

Intrinsic utility's additivity justifies taking nonproductive conjunction of BITs' objects as its concatenation operation. The choice of concatenation operation is not arbitrary. Probability theory, for example, may adopt disjunction of exclusive propositions as a concatenation operation and make probability an additive function by definition. However, if probability is degree of belief, then its additivity needs justification, and taking disjunction as a concatenation operation similarly needs justification. Carnap ([1950] 1962: 164–65) makes this point for probability taken as strength of evidence, and the point applies to intrinsic utility taken as degree of intrinsic desire. Intrinsic utility's additivity is not empirical as for weight, but normative as for probability taken as an ideal agent's rational degree of belief, and so needs a justification. Section 3.5 shows that intrinsic utility's additivity is normative rather than an immediate implication of intrinsic utility's definition. A conventional choice of a concatenation operation does not ground its additivity. The representation of intrinsic attitudes does not stipulate a concatenation operation. Degrees of intrinsic desire depend on intensities of intrinsic desires and aversions. A norm for them requires nonproductive conjunction to be a concatenation operation for BITs' objects.

Consider adding a basic intrinsic desire's realization to a world to transform it into another world. A rational ideal agent has an intrinsic preference for the new world. Using conjunction of BITs' objects as concatenation makes a representation of intrinsic preferences assign a greater degree of intrinsic desire to the new world, and any justified concatenation operation must do the same. Norms of intrinsic preference support nonproductive conjunction's being a concatenation operation for BITs' objects.

Suppose that two objects of basic intrinsic desires are equally desired intrinsically. Then, they ought to be intrinsically desired twice as much conjoined as one alone, given their conjunction's nonproductivity. The additivity of their conjunction is a norm. In an experimental situation, if an agent's degree of intrinsic desire for one pleasure is 1, and her degree of intrinsic desire for another pleasure is also 1, then assuming that these intrinsic desires are basic and that their objects are nonproductive in combination, her degree of intrinsic desire for the combination of pleasures is 2 if in this case she imitates a rational ideal agent. The norm of additivity justifies nonproductive conjunction of BITs' objects as a concatenation operation for measurement of degrees of intrinsic desire.

This chapter introduced basic intrinsic attitudes and intrinsic utility. Then, it argued for intrinsic utility's separability by arguing for its additivity. It also argued for intrinsic utility's separability directly. The independence of basic intrinsic attitudes supports the independence of their utilities. Given a partition of BITs a world realizes, the intrinsic utility of realizing a component of the partition is independent of the intrinsic utilities of the other components.

This chapter's account of intrinsic utility and intrinsic-utility analysis grounds separation of a world's utility into evaluations of the world's parts. Chapters 4–6 apply separability of a world part from its complement, as in Section 3.7, to specific divisions of a world into parts. They separate the utility of an option's world into evaluations of the option's past and its future, evaluations of the option's region of influence and the region's complement, and evaluations of the option's consequences and other events. In a decision problem, these separations license an evaluation of an option that concentrates on part of the option's world, and evaluating an option according to a part of its world simplifies choice.

CHAPTER 4

Temporal utility

When choosing seeds to plant, a gardener looks to the future. Which seeds will yield hardy plants with tasty and nutritious produce? Although the gardener uses past experiences to guide his selection of seeds, his evaluation of his options concentrates on evaluations of future events and does not include an evaluation of past events. Because all possible futures share those past events, their evaluation does not affect comparisons of options.

May an evaluation of options skip evaluation of the past? The past sets the stage for the future. May it be relevant? Focusing on the future needs a justification. This chapter justifies a method of looking forward. The method divides the utility of an option's world along the dimension of time. It evaluates separately the world's past and the world's future. The evaluation introduces temporal utility, and a form of utility analysis employing it, temporal-utility analysis, that simplifies utility comparisons of options. Initial applications assume omniscient agents. Later applications discharge that assumption.[1]

Temporal-utility analysis justifies evaluation of options according to their futures and so justifies a method of simplifying choices. The last sections of the chapter show how this common simplification of choices supports versions of traditional analyses of the utility of a person's life and traditional analyses of investments according to risk and return. Some traditional decision procedures, made precise, maximize temporal utility.

4.1 The past's influence

We put aside the past when deliberating. Indeed, the adage to ignore sunk costs requires focusing on the future. Is the common, recommended practice justified? Is it ever reasonable to act now so that some past act

[1] I am grateful to participants in the 2009 Kline conference on Causation, Time, and Choice, especially José Luis Bermúdez, for perceptive comments on my presentation of temporal utility.

will not have been in vain? May a government reasonably insist on a favorable armistice so that its soldiers' sacrifices will not have been in vain? Although a choice cannot change the past and can influence only the future, it can make the future such that past sacrifices will not have been in vain. The principle to ignore sunk costs should not bar considering relevant aspects of the past. The past may influence the value of future events.

Suppose that a college has a tradition of holding a graduation ceremony to mark the successful conclusion of students' college careers. Holding the ceremony in the current year has value that derives from the ceremony's continuing a tradition. If an evaluation of the ceremony ignores the past, it may attribute to the current year's ceremony a lower value than if the evaluation considers the past. Consequently, deliberations that ignore the past may rank conducting the ceremony lower than not conducting it, although they would reverse the ranking if they considered the past. Putting aside the past threatens to distort the options' ranking.

Care in characterizing an option's future solves the problems facing an evaluation of options according to their futures. For simplicity, an option's evaluation may evaluate the option's future without evaluating the option's past. The future (taken to include the present) may limit the scope of the option's evaluation. Any effect the past has on the future's value arises from the past's effect on future events. The value of future events that the past affects influences the future's evaluation by a rational ideal agent. The past's influence on the value of an option's future works through an appropriate characterization of the option's future. All options share the past, and the past's value does not affect options' comparisons provided that a characterization of the option's future includes the past's influence on the option's future. An option's simplified evaluation considers the past's effect on future events but evaluates only the option's future. In the example about graduation, subtracting the past's value from the utility of a world with the current year's ceremony and also subtracting it from the utility of a world without the current year's ceremony yields evaluations of the two options that agree with a comparison of the utilities of the options' worlds.

Subtracting the past's value from the utility of each option's world to obtain the value of the option's future assumes a division of a world's utility into the values of its past and future. Success depends on an appropriate division and evaluation of parts. This chapter introduces a special type of utility it calls temporal utility and shows that comparisons of futures with respect to temporal utility match comparisons of worlds with respect to comprehensive utility. The argument for the separation shows that utility-holism does not govern a world's temporal utility given the world's temporal division into parts, just as Chapter 2 shows that utility-holism does not govern an option's comprehensive utility given the option's division according to possibilities.

4.2   Kinds of utility

Propositions represent the options in a decision problem, that is, possible decisions. The propositions yield worlds with a common past. To facilitate utility comparisons of options, a type of utility, temporal utility, permits putting aside the past. Temporal utility reduces the scope of an option's evaluation. In decision problems, it evaluates not an option's world but an option's future, taken to include the option's realization and all events following the option's realization. Because the option's future includes the option itself, it includes the present resolution of the decision problem and amounts to the option's nonpast. To accompany the new type of utility, Section 4.7 advances a method of analyzing it that relies on utility theory's standard idealizations about agents and their situations, such as agents' knowledge of a priori truths.

Types of utility differ in their methods of evaluating a proposition. Comprehensive utility is the most common type and is usually called utility *tout court*. To process all considerations, it evaluates a proposition by evaluating the possible world that would be realized if the proposition were realized, the proposition's world. An option o's comprehensive utility $U(o)$ evaluates o's world, that is, the possible world that would be realized if o were realized. Instead of comprehensive utility, temporal-utility analysis needs a type of utility that evaluates just an option's future. It uses temporal utility, which has narrower evaluative scope than has comprehensive utility.

An analysis of an option's utility into the utilities of its past and its future cannot use a type of utility that simply looks forward and when applied to the past evaluates the past and its future. This type of utility's application to the past evaluates a world and not just the past. The analysis needs an evaluation of the past's period of realization, a variable whose values depend on the past realized. So *temporal utility*, instead of evaluating an event's future, evaluates an event's period. To evaluate an option's past and its future, two temporally extended events, it evaluates the periods of an option's past and its future and combines these evaluations to evaluate the duration of the option's world. Temporal utility evaluates an option's future by adjusting evaluation's scope so that it looks exclusively at a temporal period starting now and extending into the future. It evaluates a proposition's period of realization. Letting TU stand for temporal utility and $F[o]$ stand for an option o's future, that is, o's realization and the events that would follow o's realization if o were realized, $TU(F[o])$ evaluates the period of o's future and does this by evaluating events in the period. For example, a walk's evaluation uses the temporal utility of the period including the walk and its sequel, and not the temporal utility of only the walk's period. To spotlight the future, temporal utility evaluates an option's future instead of an option's world, and to evaluate an option's future, it evaluates the period of the option's future.

4.3   Basic intrinsic attitudes

Broome (1991: 25, 29) has reservations about evaluating an option by evaluating its world's temporal intervals because he doubts that parts of an option's utility may be allocated to times without omission or double counting. He suspects that the separation loses sight of values such as longevity; the value of a long life is not just the sum of the values of its years. Broome's points show that a temporal division of utility must be circumspect. It must not omit relevant considerations. Putting aside the past must not ignore the effect that the past may have on the future's utility.

Also, taking account of the past's effect on the future's utility must not count irrelevant considerations or double count considerations. Suppose that the past realizes the proposition that p, and that the future realizes the proposition that q and also the proposition that $(p \& q)$. To evaluate the future, temporal utility should count q, if it is relevant, but neither p nor $(p \& q)$, because p is a past event and counting $(p \& q)$ along with q double counts q.

Temporal-utility analysis uses basic intrinsic attitudes (BITs) and the intrinsic utilities of their realizations. Using BITs to formulate temporal-utility analysis handles longevity, tradition, and other considerations that attach to extended periods rather than to their parts, and it prevents double counting considerations. Because it rests on BITs, temporal-utility analysis is sensitive to complementarity of events in an extended period and does not double count considerations. It assigns each BIT that a world realizes to the option's past or future, and not to both. For all worlds that options realize, temporal-utility analysis lops off the BITs their common past realizes. Its evaluation of options' futures makes a clean break with the past, so that no BIT's realization straddles the divide between past and future and thus belongs to neither or belongs to both.

Suppose that an immigrant to the United States has the basic intrinsic desire to visit all fifty states. She has already visited forty states and considers a road trip that will take her through the remaining ten states. Realization of her desire straddles the present. However, a convention of analysis may assign its realization to the future because its completion lies in the future. This chapter shows that temporal-utility analysis, given simple conventions for assigning realizations of BITs to periods, accurately analyzes a world's utility.

To prevent omission, the analysis counts realizations of BITs that realizations of combinations of BITs entail, such as the constancy of pleasure that a combination of two equally intense successive pleasures generates. To prevent double counting, it counts only realizations of BITs. For example, suppose that the past realizes accomplishment and the future realizes wisdom and so also the combination of accomplishment and wisdom. Evaluation of the future counts wisdom but not the combination of accomplishment and wisdom, although the future realizes the combination, because the intrinsic desire for the combination is not basic.

Some theorists worry that evaluating wholes by evaluating their parts ignores the parts' order. Taking beginning logic before advanced logic is better than taking the courses in the reverse order. Does temporal-utility analysis miss this point? Does it propose that the value of a sequence of courses is a sum of the value of each course considered in isolation? No, temporal-utility analysis acknowledges value accruing because of the temporal order of events. It evaluates a sequence of events by evaluating the events as they arise in the sequence and not by evaluating them in isolation, that is, without regard to their circumstances.

Section 3.7 describes a case in which a traveler's voyage to Venice satisfies a basic intrinsic desire just in case she has not already visited Venice. Does temporal utility simplify an option's evaluation if a future's temporal utility depends on such events in the past? Yes, temporal utility simplifies an option's evaluation given an identification of the BITs its future realizes. In many cases, the identification is not complex. For example, in some situations a person choosing between two tasteful colors for a new sofa may know that nothing in the past matters.

4.4 Changing utility assignments

Because an agent's BITs may change, an event during an interval that is a pro according to BITs at the interval's start may be a con according to BITs at the interval's end. What utility function accurately assesses events during the interval? When BITs change over time, identifying an appropriate utility function is challenging, as Gibbard (1986: sec. 4) observes. Even a utility function that is constant during the interval may be inappropriate. Suppose that a gambler generally does not approve of his desire to gamble. He wants to eradicate it. Despite his higher-order desire, he occasionally gambles and while gambling loses his higher-order aversion and thoroughly enjoys gambling, although afterward he is remorseful. Consider the utility for him of a night's poker game. What utility function settles whether the poker game is a pro or a con? It is tempting to use the utility function that obtains during the poker game, assuming it is stable. The poker game has high utility according to it. However, that utility function ignores the significance of desires not to have certain desires. Intuitively, the game has low utility because the gambler's prevailing utility function discounts the pleasure of gambling.

This chapter bypasses the problem of evaluating utility through time. It makes temporal utility relative to a utility assignment at a time and idealizes to cases in which this utility assignment is fully rational. Despite temporal utility's relativity to time, it still regulates decisions. It is rational for an ideal agent to decide according to her current comprehensive utility assignment given its full rationality.[2] Hence, given that temporal and comprehensive utility order options the same way, as Section 4.10 argues, it is rational for her to decide according to her current temporal-utility assignment. Temporal utility relativized to a decision problem's time evaluates options, and a decision principle using temporal-utility analysis recommends an option that maximizes temporal utility so relativized. The idealization that the current utility function is fully rational prevents shortsightedness. Being fully rational, it rests on rational goals and endorses rational means of pursuing the goals. It reflects rational attitudes toward satisfaction of future desires. Given the idealization, the decision principle does not sanction, for instance, pure time-preference, that is, discounting future desires just because they are future desires. If an ideal agent is fully rational, she has higher-order desires for the satisfaction of future desires. These higher-order desires influence her current comprehensive-utility assignment and also her current temporal-utility assignment.[3]

---

[2] Weirich (1981) and Fuchs (1985) discuss this proposal. It accords with the present-aim theory of rationality in Parfit (1984: chap. 6, sec. 45), in particular, the critical version of the present-aim theory.

[3] An ideal agent ought to have a desire, not necessarily an on-balance desire, to satisfy future desires. Unless another attitude, such as an aversion to satisfaction of a particular future desire, opposes the desire to satisfy future desires, discounting future desires is irrational, as Weirich (1981) argues.

4.5   Times of realization

Temporal utility presumes an assignment of utilities to times. It entertains new locations for utility but not new sources of utility. All considerations bearing on utility that it entertains reduce to considerations concerning basic intrinsic desires and aversions. Because the utility accruing from a BIT's realization has a time, temporal-utility analysis may treat the time rather than the attitude. Nonetheless, the utility's time is just a location and not a source of utility. Temporal utility derives from the intrinsic utilities of realizations of BITs.

Everyday deliberations about decisions often use temporal utility. However, a precise presentation of this type of utility confronts puzzles concerning a proposition's time of realization. This chapter dodges these puzzles by stipulating *times of realization*; it adopts conventions for assigning to times the events a world realizes. The conventions are not contrary to intuition and just supplement intuition in cases in which it is silent. Because of them, temporal-utility analysis neither omits nor double counts realizations of BITs and assists evaluation of options without giving any option an advantage. Nonconventional features of options' evaluations represent an agent's reasons for options, and these nonconventional features yield the options' ranking. Temporal utility assessed using the conventions of location yields reliable evaluations of the options in a decision problem, as Section 4.10 shows.

A maximal consistent proposition represents a world. The realization of a nonmaximal proposition that does not represent a world may occur at a time during the world's duration. This chapter assumes that time's structure has a linear representation, and it assigns a nonmaximal proposition's realization to a position in time's structure or to a point in time's linear representation. A BIT's realization, a realization of the BIT's object (typically a nonmaximal proposition), is an event that by convention occurs at a particular time.

Alternative, principled ways of locating events, when available, fit the general framework of temporal-utility analysis. Because the chapter presents a general account of temporal-utility analysis, it presents a general method of assigning times of realization to events rather than leaving the assignment open for particular applications of temporal-utility analysis to settle according to convenience. However, particular applications may adjust the assignment.

Temporal-utility analysis justifies deciding between options according to their futures' temporal utilities rather than their outcomes' comprehensive utilities. Temporal-utility analysis takes an option's future to be the option itself together with all simultaneous and subsequent events. This is a temporal segment of a possible world. Because an option is a possible decision, the temporal location of its realization is the moment of decision. If options were acts of all types, some puzzles would arise about their temporal locations. Take becoming a philosopher. Does it happen when a scholar declares a major in philosophy, enters a graduate program in philosophy, or accepts a post in a philosophy department? It is hard to say. Similar puzzles arise for events besides acts. Take a storm's arrival. If you decide to grab your umbrella as you go out because some drops are freckling the sidewalk, and later the rain becomes a downpour, is the storm's arrival part of your decision's future, or did it precede your decision? It is hard to say. Temporal-utility analysis makes assumptions to settle such questions. It temporally locates realizations of basic intrinsic attitudes, conforming to intuition as much as possible and, for precision, settling questions that intuition leaves open.

Intrinsic-utility analysis needs a way of assigning utilities to times. Utility attaches primarily to options, outcomes, and other objects with propositional representations. When a utility is assigned to a time, the assignment must be indirect. Intuitively, utility should be assigned to a time according to the utility of realizing the desires and aversions realized at the time. More precisely, to avoid double counting, the assignment should go by the intrinsic utility of realizing the basic intrinsic desires and aversions realized at the time. If an agent has basic intrinsic desires that p and that q, and p is realized at time t_1 and q is realized at a later time t_2, then both the nonbasic intrinsic desire that $(p \ \& \ q)$ and the basic intrinsic desire that q are realized at t_2. Temporal utility nonetheless should count only realization of the desire that q in assigning a utility to t_2 so that the temporal utility of a period containing both t_1 and t_2 does not count twice realization of the desire that p.

To obtain a suitable assignment of utilities to times, temporal-utility analysis adopts a time of realization for each BIT realized whatever proposition forms its object. Each BIT realized is assigned one time, and no BIT realized is assigned more than one time. Intrinsic-utility analysis supports temporal-utility analysis only if the assignment of utilities to times separates considerations this way. If a desire for pleasure now is satisfied, its time of satisfaction is clearly now. That is, the desire is satisfied now if and only if the agent has pleasure now. Less simple desires raise complications. Suppose that an agent has a desire to leave her fortune to charity, and suppose that her fortune goes to charity after she dies. To what time should satisfaction of her desire be assigned? No time during her life seems right, yet if satisfaction of the desire is assigned to a time after her death, its satisfaction does not boost an evaluation of her life.

The first step toward an assignment of times of realization to BITs realized considers how to assign realizations of BITs to intervals. The second step considers how to assign realizations of BITs to moments. The retreat to intervals helps intuition with cases, such as the satisfaction of a basic intrinsic desire to swim across the English Channel; satisfaction of the desire requires that events go a certain way during an interval, not just at a moment. Still, some difficulties remain. Suppose that a cinematographer wants to make a movie, and suppose that she succeeds after a series of setbacks that interrupt her work. Should satisfaction of the desire be assigned to a noncontinuous interval? Next, suppose that an agent wants to have many friends and succeeds. Because "many" and "friend" are not precise, the object of the desire is a vague proposition. It is hard to select a precise interval to which to assign the desire's satisfaction. The agent's whole life is big enough, but too big. Also, consider a desire for lasting peace. If war never breaks out during an infinitely long future, the desire is satisfied. Is there any finite interval to which its satisfaction may be assigned? Some events occur during a finite interval that lacks an endpoint. A sprinter prepares for a race right up to the moment the starting gun fires and the race begins. Her preparations have no last moment but have the moment of the race's start as a least upper bound. Do her preparations lack a moment of completion?

Rather than resolve these issues, the chapter just assumes that for each proposition a world realizes, an interval exists containing just the moments with events contributing to the proposition's realization. The chapter does not specify the interval in every case but assumes the interval's existence, at least according to conventions resolving vagueness and the like. It takes this interval as the proposition's interval of realization in the world.

A conventional assignment of intervals to propositions a world realizes does not jeopardize the accuracy of temporal-utility analysis. However, temporal utility's assignment, to minimize practical problems, follows clear intuitions and stipulates only when intuitions are unclear. An intuitive assignment simplifies applications of temporal-utility analysis. It puts an agent in a position to know the temporal utility of a proposition in a world by appeal to intuitive rather than conventional criteria. In any case, the chapter's idealizations facilitate applications. Ideal agents know the assignment of intervals to propositions. Knowledge of the interval temporal utility assigns to a proposition in each possible world that realizes the proposition is a priori because it depends only on understanding temporal utility's conventions.[4]

Temporal utility assigns times of realization to BITs with respect to a current decision problem. It assigns a moment of realization to a BIT an option's world realizes. First, it assigns to the BIT's object an interval during which it is realized. Then, as the BIT's time of realization, it assigns the last time in that interval if the interval has a last time. Otherwise, if the interval contains a future moment, it assigns to the BIT an arbitrary future moment in its interval; if its interval contains no future moment, it assigns an arbitrary time in the interval. Thus, if the BIT's interval of realization spans the present and the interval has no last time, the BIT's time of realization is an arbitrary future time in the interval. If a BIT's interval of realization is open and has the present as its least upper bound, its time of realization is nonetheless in the past because its interval is completely in the past.[5]

Temporal utility assigns the realization of a BIT to the last time in the interval assigned to the proposition that is the BIT's object, assuming that the interval has a last time. This is the time that completes the BIT's realization. Thus, a person's basic intrinsic desire that her fortune go to charity, given that it is realized, is assigned to the time at which the money is bequeathed, although that time is after her death, so that the temporal utility of her life is not enhanced by the basic intrinsic desire's fulfillment. If the interval assigned to a proposition has no last moment, for example, if the proposition states a universal generalization about all times in a world of infinite duration, then the realization of a basic intrinsic desire that the proposition hold is assigned to a time in the interval associated with the proposition. The time is the present or in the future if events contributing to the event's occurrence are at the present or in the future. By convention, temporal utility assigns to the future any event that is partly in the future.

Temporal utilities, although they rest on conventions, may accurately compare the options in an agent's decision problem. This chapter compares options according to the temporal utilities of their futures rather than according to the temporal utilities of the agent's life given each option. Their futures include events that occur after the agent's death. Even if the

---

[4] If full understanding of a proposition requires knowledge of certain conventions, and the proposition's truth follows from these conventions, then knowledge of the proposition is a priori. For example, knowledge of chess piece positions that count as checkmate is a priori because full understanding of checkmate requires knowledge of chess's conventions, and they settle what counts as checkmate. Full understanding of temporal utility requires knowing its conventions, which settle its interval assignments. So, knowledge of temporal utility's interval assignments is a priori.

[5] Thanks to Joshua Smart for helpful points about times of realization.

temporal utility of a life depends only on events occurring during the life, the temporal utility of an option's future includes relevant events after the agent's death, such as honoring the agent's will.

For BITs an option's world realizes, temporal utility locates each BIT's realization at a time so that a division according to time does not omit or double count any BIT's realization. An analysis can obtain the intrinsic utility of a world's segment by adding the intrinsic utilities of BITs realized in it according to the analysis. The method of separating realizations of BITs in a temporal interval of the world and its complement may arbitrarily assign a BIT's realization to the interval or its complement; any method of separating BITs realized in an option's world works, although practicality advises conforming to firm intuitions.

Temporal-utility analysis divides a world's temporal utility into the temporal utilities of the past and the future. The division need not derive from division of a world's temporal utility into an arbitrary number of parts. The chapter does not need temporal-utility analysis of an arbitrary proposition, but only the temporal utilities of a world, its past, and its future. Although the chapter may omit temporal-utility analysis of an arbitrary proposition's temporal utility, it includes a general analysis to explain the analysis of a world's temporal utility into its past's and its future's temporal utilities. Temporal-utility analysis of an arbitrary proposition clarifies the past's and the future's temporal utilities.

Temporal utility assigns realization times to BITs, but temporal-utility analysis of a world's temporal utility using the world's past and future may simply assign to the future any event that overlaps the future, without reliance on realization times. The objective is to prevent omission and double counting by assigning each BIT realized to just one element of a partition of the world's interval. Division of a world's temporal utility into temporal utilities of its past and its future does not need a general assignment of realizations of BITs to times good for all applications of temporal-utility analysis. Temporal-utility analysis applied to arbitrary partitions of arbitrary intervals needs times of realization, but temporal-utility analysis of a world divided into past and future need only assign to the future events that have parts in the future.

4.6   Temporal utility given full information

Temporal utility allocates to times realizations of desires and aversions. Then, it evaluates a temporal interval by evaluating the desires and aversions the interval realizes. Given full information, an agent may compute an interval's temporal utility.

Suppose that the desire d_1 is realized during the interval $[t_1, t_2)$, and the desire d_2 is realized during the immediately succeeding interval $[t_2, t_3]$. No other desires or aversions are realized during either the first or second interval. Is the temporal utility accruing during the compound interval $[t_1, t_3]$ the sum of the utilities accruing during $[t_1, t_2)$ and during $[t_2, t_3]$? Perhaps not. The desires d_1 and d_2 may be for pleasures. If the intensity of those pleasures is the same, and the agent wants pleasure of a constant intensity during $[t_1, t_3]$, then the realization of d_1 and d_2 realizes the desire for constancy. Because of complementarity, the temporal utility accruing during $[t_1, t_3]$ is greater than the sum of the temporal utility accruing during $[t_1, t_2)$ and the temporal utility accruing during $[t_2, t_3]$.

So that temporal utility is additive, its definition makes it rest on basic intrinsic attitudes. Using them eliminates the dangers of omission and double counting. As Section 3.5 shows, intrinsic utility is additive: the intrinsic utility of a conjunction of BITs' objects equals the sum of the intrinsic utilities of the BITs' objects that the conjunction entails. If the desires for pleasure and its constancy are the only BITs, then letting d_1 and d_2 stand, respectively, for the objects of the desires for pleasure, $IU(d_1 \& d_2) = IU(d_1) + IU(d_2) + IU(\text{constancy})$ because the combination of the pleasures entails constancy. To evaluate an interval, temporal utility attends to all and only the BITs the interval realizes.

The constancy of pleasure during the two intervals, according to conventions of location, occurs at the endpoint of the union of the intervals. Although the locations of realizations of BITs, being settled by convention, do not explain a period's temporal utility, temporal-utility analysis accurately obtains a period's temporal utility from the temporal utilities of subperiods.

Given an assignment of realizations of propositions to intervals, temporal utility constructs an assignment of realizations of BITs to times, as in Section 4.5. Then, it assigns utilities to times in a world according to the BITs realized at those times. The total utility assigned to a time is the sum of the intrinsic utilities of the objects of the BITs realized at the time. To obtain the utility of an interval, temporal utility adds the utilities of the times contained in the interval. The *temporal utility*, *TU*, of any proposition a world realizes is the temporal utility of the interval assigned to the proposition in the world. Given complete information, this is the sum of the intrinsic utilities of the objects of the BITs with times of realization in the interval. In general, given complete information, for any proposition p,

$$TU(p) = \sum_i IU(\text{BIT}_i)$$

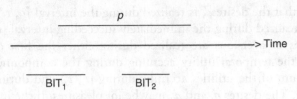

Figure 4.1 A proposition's temporal utility

where BIT_i ranges over the objects of the BITs with times of realization in the interval assigned to p in the world that would be realized if p were realized, that is, the objects of BITs with times of realization in p's interval in p's world. This is a definition of temporal utility, not a normative principle to defend.

Figure 4.1's illustration has a time line for the world that would be realized if p were realized. The interval assigned to p is depicted above the time line. Two basic intrinsic attitudes BIT_1 and BIT_2 have times of realization in that interval. The intervals of the propositions that are their objects are depicted below the time line. The interval assigned to p does not contain both intervals. Nonetheless, the right end-points of the intervals are in the interval assigned to p. Assuming that the only BITs with times of realization in the interval assigned to p are BIT_1 and BIT_2, the temporal utility of p is the sum of the intrinsic utilities of realizing these BITs. That is, $TU(p) = IU(BIT_1) + IU(BIT_2)$, with BIT_1 and BIT_2 in this formula standing for the BITs' objects.

A proposition represents an *event*, as this chapter understands events. Hence, utilities attaching to events attach to propositions. A proposition's temporal utility evaluates directly the events in a proposition's interval. It yields indirectly an evaluation of the proposition's interval using appraisals of the events in the interval. Temporal-utility analysis uses evaluations of propositions characterizing intervals to divide temporal utilities.

The temporal utility of realizing a BIT may differ from the intrinsic utility of realizing the BIT because its interval of realization may contain the realizations of additional BITs. All BITs assigned to its interval are tallied when computing its realization's temporal utility. The temporal utility evaluates events that occur contemporaneously with the BIT's realization, as opposed to just events its realization entails.

Given complete information, a proposition's temporal utility equals the sum of the intrinsic utilities of objects of BITs with times of realization in the proposition's interval. A conjunction of objects of BITs represents the set of BITs. According to intrinsic utility's additivity, its intrinsic utility equals the sum of the intrinsic utilities of the objects of BITs it entails. Therefore, for a nonproductive conjunction of BITs' objects that characterizes an interval, its temporal and its intrinsic utilities are identical. For example, a conjunction of BITs' objects that characterizes a world's duration has identical temporal and intrinsic utilities.

Some comprehensive utilities equal temporal utilities given complete information. Given complete information, the comprehensive utility of a proposition p, $U(p)$, equals the intrinsic utility of the world that would be realized if the proposition were realized, that is, p's world, $W[p]$. By intrinsic-utility analysis, the possible world's intrinsic utility equals the sum of the intrinsic utilities of the objects of the BITs the world realizes. $TU(p)$ need not equal $U(p)$ given complete information because it has narrow scope. It evaluates only the interval of p's realization in p's world. It does not evaluate p's entire world. However, the temporal utility of p's world, $TU(W[p])$, equals the sum of the intrinsic utilities of the objects of the BITs realized at times in the world's interval of realization, namely, the world's duration. Every BIT realized in the world is realized at a time in the world's interval. Thus, the sum of the intrinsic utilities of the objects of the BITs realized at times in the world's interval equals the sum of the intrinsic utilities of the objects of the BITs the world realizes. Therefore, given complete information, the temporal utility of p's world equals p's comprehensive utility. That is,

$$U(p) = TU(W[p])$$

This equality shows how temporal utilities yield comprehensive utilities.

4.7 Temporal-utility analysis

Temporal-utility analysis breaks down a proposition's temporal utility using temporal intervals. It assists evaluation of propositions. This section presents temporal-utility analysis for agents who are certain of the possible world that would obtain if a proposition were realized. The next section adjusts the form of analysis to accommodate uncertainty.

A business may evaluate years, and an employee may evaluate days. An evaluation may break a period into smaller periods to evaluate. Evaluation of a proposition may consider it with respect to a partition of its interval. The temporal utility of a proposition p, $TU(p)$, may be analyzed temporally. It may be broken down into the temporal utilities of propositions representing the members of a partition of the interval assigned to p in p's world.

An analysis of a world's temporal utility into past and future components uses such a division.

Given complete information, an agent is certain of the interval assigned to p in p's world. Consider a finite partition of p's interval. Let p_i be a proposition characterizing the ith element of the partition (it may say that time passes from the subinterval's start to its end). Given complete information, the agent is certain of the BITs realized at times in the subintervals and is certain that they are the BITs realized at times in p's interval. Thus, given complete information, this form of temporal-utility analysis emerges:

$$TU(p) = \sum\nolimits_i TU(p_i)$$

where p_i ranges over propositions characterizing intervals that partition p's interval of realization in p's world. This identity holds because each BIT realized in p's interval is assigned to a unique time in the interval. Hence, adding the intrinsic utilities of the objects of the BITs realized at times in the subintervals yields the same result as adding the intrinsic utilities of the objects of the BITs realized at times in the entire interval.

Figure 4.2 provides an illustration. It is the same as Figure 4.1 except that it partitions p's interval into two intervals characterized by p_1 and p_2. BIT$_1$ is assigned to p_1's interval because it is realized at a time in p_1's interval, and BIT$_2$ is assigned to p_2's interval because it is realized at a time in p_2's interval. BIT$_1$ and BIT$_2$ are the only BITs realized at times in p's interval, and each is realized at a time in exactly one of the intervals assigned to p_1 and to p_2. Hence, $TU(p) = TU(p_1) + TU(p_2)$. This follows from features of time and temporal utility and is not a definition.

Given complete information, according to temporal-utility analysis and the definition of temporal utility, the temporal utility of a world w equals the sum of the intrinsic utilities of the objects of BITs realized at times in the intervals of propositions w_j characterizing the elements of an n-fold partition of the world's duration. That is, $TU(w) = \sum_j TU(w_j) = \sum_j \sum_k IU(\text{BIT}_{jk})$, where BIT$_{jk}$ ranges over the objects of BITs realized at times in partition element j according to w. $TU(w)$ calculated this way has the same value as

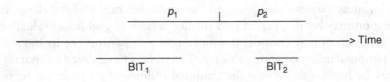

Figure 4.2 Temporal analysis of $TU(p)$

$U(w)$ calculated using intrinsic-utility analysis. As intrinsic-utility analysis shows, $U(w)$ equals $IU(w)$, which is the sum of the intrinsic utilities of the objects of BITs realized in w. The formula for $TU(w)$ just provides another way of summing those intrinsic utilities. Temporal- and intrinsic-utility analyses offer consistent methods of calculating $U(w)$, and the chapter's appendix shows the consistency of other methods of calculating utilities.

4.8   Temporal utility given incomplete information

This section discharges the assumption of certainty about the possible world that would obtain if a proposition were realized. It generalizes temporal-utility analysis to handle incomplete information. It also explains temporal utility's relation to other types of utility and temporal utility's contribution to decision theory.

Incomplete information affects temporal utility in three ways. First, an agent may not know whether the proposition being evaluated is true and so may not know whether to evaluate it with respect to actual or hypothetical circumstances. For example, an agent may not know whether her lottery ticket's winning the prize occurs in the actual world or in a nonactual world. Should she evaluate the proposition that it wins in the actual world or in a nonactual world? Second, even if the agent knows the proposition is true, she may not know the way the proposition is realized and so may not know the interval assigned to the proposition. For example, if the proposition is a disjunction, its interval may vary according to whether it is true because of the first, second, or both disjuncts. Also, if the proposition does not specify a time of realization, as the proposition that the weather improves, the interval assigned to the proposition's realization varies according to the circumstances of its realization. In such cases, ignorance of the circumstances of a proposition's realization may cause ignorance of its interval of realization. Third, the agent may not know the basic intrinsic attitudes realized in the proposition's interval of realization. For instance, take the proposition that she visits the Louvre today. She may be uncertain whether her visit would satisfy her basic intrinsic desire to see the Mona Lisa because she thinks the painting might be undergoing restoration. This uncertainty is about the events occurring in the proposition's interval of realization, rather than about the moments the interval comprises.

Expected-utility analysis separates a proposition's utility into the utilities of the prospects for its various possible outcomes, that is, the probability-utility products for the possible outcomes. It takes a proposition's utility to equal its outcome's expected utility. Temporal-utility analysis accommodates uncertainty by adopting expected-utility analysis's method. It obtains the temporal utility of a proposition from the expected value of the proposition's temporal utility with respect to possible resolutions of the relevant uncertainty. Resolutions more coarse-grained than possible worlds may work (say, nonmaximal propositions that sets of possible worlds represent), but possible worlds work in every case, so temporal-utility analysis may rely on them. Accordingly, it takes the temporal utility of a proposition as the proposition's expected temporal utility with respect to the possible worlds that might be realized if the proposition were realized.

Consequently, a good procedure for obtaining a proposition's temporal utility starts by identifying the possible worlds that might be true if the proposition were true. Each possible world realizing the proposition resolves uncertainty about the circumstances of the proposition's realization and therefore about the proposition's interval of realization. It also resolves uncertainty about the BITs realized during the proposition's interval of realization. Next, the procedure obtains the temporal utility of the proposition with respect to each of the possible worlds. Finally, it computes the proposition's temporal utility as the probability-weighted average of its temporal utility in the worlds realizing it. This general definition of temporal utility summarizes the procedure:

$$TU(p) = \sum_i P(w_i \text{ given } p) TU(p \text{ given } w_i)$$

where w_i ranges over possible worlds where p is true (other worlds have zero probability given p so that the probability-utility products for them do not affect a summation over all worlds). The conditional probability $P(w_i$ given $p)$ is not the usual ratio of nonconditional probabilities. It is the probability of w_i under the supposition that p, as Section 2.3 explains.

In the formula for $TU(p)$, $TU(p$ given $w_i)$ equals $TU(p$ given $(p \,\square\!\!\rightarrow w_i))$, for each world w_i. The proposition p is true in w_i, so the supposition that $(p \,\square\!\!\rightarrow w_i)$ reduces to the supposition that w_i. Hence, the definition of temporal utility implies for conditional temporal utility an analog of the interpretation of conditional utility that Section 2.5 presents.

An analysis of a proposition's temporal utility, according to an n-fold partition of the proposition's interval, proceeds as follows. Given a possible world that realizes the proposition, analyze the temporal utility of the proposition with respect to the n-fold partition of the interval assigned to it. Then, obtain the expected value of such an analysis of the temporal utility of the proposition with respect to all worlds that realize it. That is, start with the formula $TU(p) = \sum_i P(w_i$ given $p) TU(p$ given $w_i)$, where w_i ranges over possible worlds that make p true. Analyzing $TU(p$ given $w_i)$, this formula becomes

$$TU(p) = \sum{}_i P(w_i \text{ given } p)\sum{}_j TU(p_{ij})$$

where p_{ij} ranges over propositions characterizing the members of an n-fold partition of p's interval given w_i.

Temporal-utility analysis's application to decision problems requires a way of connecting temporal utility with comprehensive utility because the fundamental decision rule is to maximize comprehensive utility. Examining the nature of comprehensive utility reveals a suitable connection. $U(p)$ is an evaluation of the outcome of p's realization. $W[p]$, p's world, is the world that would be realized if p were realized. $O[p]$, p's outcome, is the proposition that $W[p]$ obtains. $W[p]$ is a maximal consistent proposition. In contrast, $O[p]$ is not maximal because $\ulcorner W[p] \urcorner$, a component of its standard expression, names but does not specify a world. Given uncertainty about p's world, $TU(O[p])$ may differ from $TU(W[p])$ because $TU(O[p])$ is a probability-weighted average of the temporal utilities of p-worlds, not the temporal utility of a single p-world. Also, $TU(W[p])$ may differ from $U(p)$. In general, the latter is a probability-weighted average of the intrinsic utilities of p-worlds, not the intrinsic utility of a single p-world. However, $TU(O[p])$ is the same probability-weighted average as $U(p)$. Hence,

$$U(p) = TU(O[p])$$

Temporal-utility analysis evaluates an option o in a decision problem using $TU(O[o])$, the temporal utility of the option's outcome. The evaluation uses $TU(O[o])$ instead of $TU(o)$ because the former appraises a whole world, as does $U(o)$, whereas $TU(o)$ evaluates only the period of o's realization. $TU(O[o])$ evaluates o's outcome comprehensively. It takes the proposition $O[o]$ as a lottery over possible worlds. So $TU(O[o])$ is the probability-weighted average of the temporal utilities of possible worlds. That is, $TU(O[o]) = \sum{}_i P(w_i \text{ given } o) TU(w_i)$, where w_i ranges over possible worlds. The sum may be restricted to o-worlds because only they have positive probabilities given o. The temporal utility of a possible world is the sum of the temporal utilities assigned to subintervals dividing the world's duration. Temporal division of a world w's temporal utility yields the equality $TU(w_i) = \sum{}_j TU(w_{ij})$, where the w_{ij}'s characterize the elements of an n-fold partition of w_i's duration. Combining the analysis of $TU(O[o])$ with this temporal division of $TU(w_i)$ yields the general form of temporal-utility analysis:

$$TU(O[o]) = \sum{}_i P(w_i \text{ given } o)\sum{}_j TU(w_{ij})$$

Assuming complete information, to obtain a proposition's temporal utility given a condition, one may conditionalize the components of Section 4.7's analysis of the proposition's temporal utility with respect to a partition of the proposition's interval of realization. Accordingly, $TU(p$ given $q) = \sum_i TU(p_i$ given $q)$, where the p_is characterize intervals that form a partition of p's interval. The proposition p_i's temporal utility given q varies according to the p-world realized given q. So, to allow for incomplete information, one may apply expected-utility analysis, taking $TU(p$ given $q)$ as the expected value of $\sum_i TU(p_i$ given $q)$ with respect to possible worlds that realize p and q. The formula for $TU(p$ given $q)$ considers only worlds that realize p and q, and not all worlds that realize q. It divides p's interval in a $(p \& q)$-world using a partition $\{p_i\}$. No partition exists if p does not obtain.

Conditional temporal-utility analysis may be used during an option o's standard expected-utility analysis to compute o's comprehensive utility given a state s, $U(o$ given $s)$. Given a state s, the temporal utility of an option's outcome equals the option's comprehensive utility: $TU(O[o]$ given $s) = U(o$ given $s)$. Methods of computing $TU(O[o]$ given $s)$ therefore provide additional methods of computing $U(o$ given $s)$. Adding temporal-utility analysis to other methods of utility analysis multiplies ways of computing a proposition's comprehensive utility. Consistency prevails because all methods of analysis reduce to Section 3.8's canonical analysis. The chapter's appendix verifies that temporal-utility analysis is consistent with other methods of utility analysis.

4.9 Futures

When applying temporal-utility analysis to decision problems, this chapter ignores an option's temporal utility because that utility evaluates only the narrow interval during which the option is realized. For breadth of evaluation, the chapter attends to the temporal utility of the option's future.

Suppose that o is an option. Let $F[o]$ stand for o's future, the world segment starting with o that would obtain if o were realized. It is a consistent proposition that is maximal with respect to events with times of realization at or after the time of o's realization. The interval for o's future is the world segment's duration. Whereas the comprehensive utility of $F[o]$ evaluates the whole world containing o's future, the temporal utility of $F[o]$ evaluates merely the events that occur in o's future.

A derivative proposition accommodates uncertainty about $F[o]$. Let $F'[o]$ be the proposition that $F[o]$ obtains. $F'[o]$, in contrast with $F[o]$, is not maximal concerning events realized at or after the time of o's realization.

Given o's adoption, $F'[o]$ is known, whereas $F[o]$ is typically unknown. Similarly, $TU(F'[o])$ is known, but $TU(F[o])$ is typically unknown.

A suitable evaluation of o computes $TU(F'[o])$, which is the expected value of $TU(F[o])$. Its formula is

$$TU(F'[o]) = \sum{}_i P(f_i \text{ given } o) \, TU(f_i)$$

where f_i ranges over o's possible futures, that is, possible denotations of $F[o]$. According to Section 4.6's definition of temporal utility, $TU(f_i) = \sum{}_j IU$ (BIT_{ij}), where BIT_{ij} ranges over the objects of BITs realized at times in f_i's interval, that is, its duration. Applying the definition to the formula for $TU(F'[o])$ yields the equation

$$TU(F'[o]) = \sum{}_i P(f_i \text{ given } o) \sum{}_j IU(\text{BIT}_{ij})$$

Applying instead a temporal-utility analysis of $TU(f_i)$, using a partition of f_i's interval, yields the equation

$$TU(F'[o]) = \sum{}_i P(f_i \text{ given } o) \sum{}_k TU(f_{ik})$$

where f_{ik} characterizes the kth member of the partition of f_i's interval. Are the two derived formulas for $TU(F'[o])$ consistent?

Using Section 4.6's definition of temporal utility for $TU(f_{ik})$, and noting that all BITs realized at times in an element of f_i's partition are realized at times in f_i's interval, the second derived formula becomes the equation

$$TU(F'[o]) = \sum{}_i P(f_i \text{ given } o) \sum{}_j IU(\text{BIT}_{ij})$$

where, as before, BIT_{ij} ranges over the objects of BITs realized at times in f_i's interval. Because the second derived formula may be transformed into the first derived formula, calculations of temporal utilities pass a consistency check of the sort the chapter's appendix conducts.

4.10   **Separability**

Intrinsic utility's additivity justifies, using methods Chapter 3 describes, temporal-utility analysis. This section examines additivity's justification of temporal-utility analysis's separation of a world's temporal utility into its past's and its future's temporal utilities. Temporal utility's separability justifies evaluating options according to their futures. Its order of options according to their futures is the same as its order of options according to their worlds, as the section shows.

Temporal utility's separability takes a world's parts as instantiations of variables. It takes the past as a variable whose value is a particular past, such as the actual world's past. Similarly, it takes the future as a variable.

Temporal utility orders the values of these variables. Given separability of the future from the past, the order of futures agrees with the order of their worlds, given any way of fixing the past. Given separability of the future's temporal utility from the past's temporal utility, the temporal-utility function for futures settles the ranking of their worlds, given any way of fixing the past's temporal utility. Before establishing this type of separability, some fine points need clarification.

Separability applies straightforwardly to goods in commodity bundles. Variables have amounts of a good as values. All combinations of commodities are possible. One glove may combine with any other glove to form a pair. Separability of temporal utility involves combinations of pasts and futures that form worlds; it entertains only combinations that are possible. Which combinations are possible is an issue in the metaphysics of possible worlds.

The theory of branching time, as expounded, for example, in Horty (2001: sec. 2.1), asserts that the past exists and the future does not yet exist. The past may be conjoined with various possible futures, none of which exists yet, and for none of which is it yet true that it will exist. The combination of the past and a possible future is a possible world. No possible world compatible with the past exists yet, except the one that ends now. If the actual world continues, it is currently incomplete. Although the actual past exists, the actual future does not yet exist – it is not even settled now that it will exist. A single past may be part of multiple worlds and have multiple futures following it. According to this theory, separability entertains possible worlds that conjoin a single past with various futures.

A traditional rival view, which Lewis (1986b: chap. 1) elaborates, asserts that all possible worlds exist now, and the possible futures they contain exist now although they are not present. Just one of the possible worlds is actual. The actual world is complete now even if it continues. It contains the past and the possible future that follows the past. That possible future is the actual future.[6] Two possible worlds may be the same up to the present and diverge in the future. According to this view also, separability entertains possible worlds that conjoin a single past with various futures.

A third view, which arises from applying Lewis's (1986b: chap. 4) counterpart theory not only to people but also to pasts, asserts that all possible worlds exist now and the actual world is complete, but no worlds

---

[6] Taylor (1962) claims that statements about the future have truth-values now. He uses this point to argue for fatalism, but the point is detachable from fatalism.

share a past. Identity of pasts entails identity of worlds because of relations between past and future that hold even in indeterministic worlds. A past identifies a whole world and is incompatible with other worlds because of its metaphysical relation to future events. If I will run a marathon in the future, then in the past I was such that I will run a marathon in the future. The actual past, which exists wholly in the actual world, belongs to the actual world and identifies that world because it belongs to no other possible world. Other possible worlds may have counterparts of the actual past, but they lack the actual past's relation to the actual future and so are not the same as it. Several worlds may share a past in the sense that events in the past unrelated to the future alike occur in the worlds' pasts. The worlds' pasts do not bear the same relations to future events, but nonetheless the worlds share wholly past events propositionally characterized. Although their pasts are not identical, they are alike in occurrences of events limited to the past. Their pasts have the same intrinsic properties. This view of possible worlds rules out separability's entertaining worlds that share the very same past. To accommodate this view, the next paragraph adjusts separability's interpretation.

Given that a world's past identifies the world, the world's past does not combine with multiple futures to form a world, as one glove may combine with multiple other gloves to form a pair. Fixing the past fixes a future and a world. Because of this disanalogy between worlds and commodity bundles, temporal utility's separability moves from metaphysically possible worlds to epistemically possible worlds. For an agent, an epistemically possible world is a maximal consistent proposition that is true for all the agent knows, assuming that knowledge implies certainty. In epistemically possible worlds, the past typically combines with various futures, even if only one combination is metaphysically possible. The values of the variable for the past are the pasts of epistemically possible worlds. Separability treats worlds for which it is epistemically possible that they share the same past and differ only in their futures. In a decision problem, attending to all epistemically possible worlds ensures a review of considerations that ignorance of the actual world makes relevant.

This chapter's points about separability, with suitable adjustments, may adopt any of these views about the compatibility of pasts and futures. Nonetheless, for the sake of definiteness and direction of additional studies of separability, the chapter takes a stand. It does not argue for its stand because nothing significant in this book depends on its stand. It simply takes the past to identify the actual future and the actual world (without assuming determinism). Hence, it takes temporal utility's separability to treat epistemically possible worlds, including those that combine the actual past with divergent epistemically possible futures. Rival views of possible worlds simplify separability by applying it to metaphysically possible worlds that combine a past with multiple futures. In garden-variety decision problems with ideal agents, the differences between epistemically and metaphysically possible worlds are not important, and a simplified version of separability suffices. The book's remainder ignores differences between the official and a simplified version of separability and speaks as though a past may combine with multiple futures.

Temporal-utility analysis splits a world's temporal utility into the past's temporal utility and the future's temporal utility. The future's temporal utility equals the world's temporal utility after subtracting the past's temporal utility. An evaluation of options using their futures' temporal utilities lops off the past's temporal utility from the temporal utility of each option's world. Comprehensive utilities of options equal temporal utilities of options' worlds. Because the temporal utility of an option's future evaluates only the option's future, not the option's entire world, it subtracts the past's temporal utility from the option's comprehensive utility. An option's temporal utility in effect changes the scale for the option's comprehensive utility. Thus, the ranking of options by temporal utilities of their futures agrees with the ranking of options by their comprehensive utilities. In a decision problem, an option has maximum comprehensive utility if and only if its future's occurrence has maximum temporal utility. Realizing an option that maximizes the temporal utility of options' futures is equivalent to maximizing the comprehensive utility of options. The support for these claims makes some assumptions.

The consistency of ranking options according to their comprehensive utilities and according to their futures' temporal utilities assumes that all the options in a decision problem have the same moment of realization. Then, the options' futures start at the same time and form world branches at a single node. The assumption of simultaneity makes the world segment prior to an option's future the same for each option. The standard stipulation that options are decisions entails the assumption. According to this stipulation, in a decision problem an agent's options are the decisions she might make at the time of the problem. These options all have the same moment of realization, namely, the time of decision. Although an agent may deliberate today about a decision she will make tomorrow, a current decision problem concerns a decision to make now.

Table 4.1 juxtaposes temporal utility's and comprehensive utility's evaluations of two options o_1 and o_2 assuming complete information. In the table, temporal utility evaluates an option o using the option's future $F[o]$.

Table 4.1 *Options' temporal and comprehensive utilities*

			Time of decision		
o_1	BIT_1	BIT_2	|		BIT_3
o_2	BIT_1	BIT_2	|		BIT_4

$TU(F[o_1]) = IU(BIT_3)$

$TU(F[o_2]) = IU(BIT_4)$

$U(o_1) = IU(BIT_1) + IU(BIT_2) + IU(BIT_3)$

$U(o_2) = IU(BIT_1) + IU(BIT_2) + IU(BIT_4)$

Comparing the options using their futures' temporal utilities yields the same result as comparing them using their comprehensive utilities. Their temporal utilities just ignore BITs the past realizes. The temporal utility of an option's future drops consideration of BITs the past realizes because each option's world realizes them. Given intrinsic utility's additivity, subtracting the intrinsic utility of the past's realization of BITs from the intrinsic utilities of the worlds that options realize does not change the options' ranking. It just subtracts common factors from a set of sums and so does not change the sums' order.

Temporal utility simplifies utility comparisons of options. According to comprehensive utility, given complete information, an option's utility is the utility of the option's world. An agent's deliberations pare down possible worlds so that they include only features that matter to the agent. Nonetheless, an option's world remains extremely complex. Using temporal utility achieves a significant simplification. Instead of considering the entire world that would result if an option were realized, an evaluation may consider just the world's temporal segment beginning at the moment of decision. That is, it may put aside the common temporal segment of the possible worlds branching at the moment of decision according to the option selected. Because the possible worlds the agent might realize have in common events in the actual world up to the time of decision, none has an advantage over the others with respect to those events. Utility comparisons of the options may put aside those past events. The temporal utility of the world segment an option starts, the option's future, puts them aside. It functions this way even given incomplete information, as Section 4.8 shows.

Given full information, efficient decision principles do not evaluate all BITs an option's world realizes but focus on the option's future and identify BITs it realizes. An evaluation using temporal utility works because a world's intrinsic utility divides into the intrinsic utilities of the world's realizations of BITs. Given the division, the past's intrinsic utility contributes in the same way to every option-world's intrinsic utility. Removing the past's contribution to possible worlds' intrinsic utilities preserves comparisons of options. Maximizing temporal utility among options' futures agrees with maximizing comprehensive utility among options' worlds.

Having clarified the type of separability to establish, this section now turns to the case for it. Support comes from Chapter 3, because intrinsic utilities define temporal utilities. The argument advances separability as a norm of rationality for ideal agents.

For rational ideal agents, the future is separable from the past. Such agents have reasons to make the order of futures agree with the order of worlds given a fixed past so that the temporal utility of worlds is strictly monotonically increasing as the temporal utility of the future increases. The separability of the variable for the future from the variable for the past follows from the additivity of the variables' values for each pair of values that make a single world. The additivity of their values' temporal utilities depends on the independence, for a world, of the past's and the future's temporal utilities. The temporal utilities are independent because the reasons for them are independent. The reasons for the past's temporal utility are the BITs the past realizes, and the reasons for the future's temporal utility are the BITs the future realizes. The reasons are independent because BITs are held independently, as Section 3.2 explains.

Because intrinsic utilities of BITs' objects define temporal utility, temporal utility's separability follows from intrinsic utility's separability. Separation of a world's intrinsic utility into intrinsic utilities of conjunctions of BITs' objects representing the past and the future that the world realizes follows from separation of a world's intrinsic utility into intrinsic utilities of realizations of sets of BITs, as Section 3.7 explains. In evaluation of worlds, the variable for the future is separable from the variable for the past, and the future's temporal utility is separable from the past's temporal utility.

The ground of temporal utility's separability, namely, intrinsic utility's additivity for the past and the future, uses separability's representations of the past and the future. Grant that a world's intrinsic utility divides into the intrinsic utilities of realizations of sets of BITs. Separability's introduction of values for variables for the past and the future identifies BITs that yield the past's intrinsic utility in the world when their objects' intrinsic utilities are summed. It does the same for the future's intrinsic utility in the world. For each BIT the world realizes, the introduction assigns the BIT's realization to either the past or the future so that adding up intrinsic utilities of objects of BITs realized in the past yields the past's intrinsic utility, and similarly for the future's intrinsic utility.

Temporal-utility analysis creates the framework for the past's and the future's separability. It attributes realizations of BITs to temporal intervals. It identifies conjunctions of objects of BITs that represent the past and the future for each world and when conjoined represent the world. Then, it applies intrinsic utility's addition principle to those conjunctions. Temporal-utility analysis identifies the intrinsic utility of the past with the intrinsic utility of the conjunction of objects of BITs that the past realizes. Similarly, it identifies the intrinsic utility of the future with the intrinsic utility of the conjunction of objects of BITs that the future realizes. The conjunction of objects of BITs realized in the past does not entail the object of any BIT realized in the future. If the combination of a past and a future generates realization of a BIT, temporal-utility analysis puts its realization into the set representing the future so that the set representing the past is the same for each option. Section 3.5 establishes that the intrinsic utility of a conjunction of objects of BITs equals the sum of the conjuncts' intrinsic utilities, provided that the conjunction has as a conjunct the object of every BIT whose realization it entails. The conjunctions representing a world's past and its future meet the condition. Hence, temporal utility's additivity follows from intrinsic utility's additivity.

Separability's expression may use conditional preferences. The order of futures given a value of the past is a conditional preference ranking for which the condition states a supposition about a variable's value. The future is separable from the past if and only if the order of futures agrees with the order of worlds no matter how the past is fixed. According to separability, the supposition about the past does not affect the conditional preference ranking of futures, which is the same as the nonconditional preference ranking of worlds with those futures. Conditional intrinsic utilities settle the conditional preferences' scope and the conditional preferences themselves. Let p_1 and f_1 be, respectively, a particular world w_1's past and future. Fixing the past's value, conditional intrinsic preference orders futures and orders worlds containing the futures. According to Section 3.3's account of conditional intrinsic utility, the identity of intrinsic and comprehensive utility for worlds, and Section 4.6's definition of temporal utility, $IU(f_1 \mid p_1) = U(p_1 \ \& \ f_1) - TU(p_1)$. $TU(f_1) = IU(f_1 \mid p_1)$, so $TU(f_1) = U(p_1 \ \& \ f_1) - TU(p_1)$; that is, $TU(f_1) = U(w_1) - TU(p_1)$. The equations hold for each world. They ensure the future's separability from the past.

Section 3.7 argues generally for separability of world parts having realizations of sets of BITs as their values. Although focusing on the future needs only the separability of the future from the past, are the variables for the past and the future mutually separable? It may seem that although the future is separable from the past, the past is not separable from the future. Among pasts of equal value, some are productive when conjoined with a future. Combine a productive past and future. Keep the future fixed, and substitute for the productive past a nonproductive past of equal value. Then, the new world fails to realize a BIT that the first world realizes, and its intrinsic utility differs from the first world's intrinsic utility. The order of pasts disagrees with the order of worlds. Hence, the past is not separable from the future, the example concludes.

This apparent counterexample to mutual separability involves some sleight of hand. In it, substituting a nonproductive past for a productive past changes the set of BITs constituting the future's value. The future's value is not fixed during the substitution. Hence, the substitution is not a counterexample to the past's separability from the future. Separability depends on a precise specification of variables and their values. Mutual separability and, in particular, the separability of the past from the future hold if the future's instantiation is a realization of a set of BITs and not just a period of time.

4.11 Holism

Objections to temporal-utility analysis allege that it neglects complementarities between the past and future. Suppose that for an artist's activities, A_1 stands for art at time t_1, and B_1 stands for banking at time t_1. Similarly, A_2 and B_2 stand, respectively, for art and banking at time t_2. The present stands between t_1 and t_2. Table 4.2 orders worlds w_1-w_4, which specify the artist's pair of activities during the times, putting the artist's top preference at the top. Additivity of temporal utilities licenses inferences from comparisons of wholes to comparisons of parts. Given a common first element, a

Table 4.2 *Complementarity*

	t_1	present	t_2
w_1	A_1		A_2
w_2	A_1		B_2
w_3	B_1		B_2
w_4	B_1		A_2

ranking of pairs implies a ranking of the pairs' second elements. Given additivity, $w_1 > w_2$ implies $A_2 > B_2$, and $w_3 > w_4$ implies $B_2 > A_2$. Because these comparisons of elements are inconsistent, additivity prohibits the table's order of worlds although rationality permits it, the objection claims.

Additivity's reply to the objection notes that the table's cells do not list all BITs realized if the order of worlds is rational for an ideal agent. Suppose that the artist has basic intrinsic desires for all the activities the table lists, and w_3 ranks higher than w_4 because of the artist's basic intrinsic desire for continuity, C. In w_1, the objects of BITs realized are A_1, A_2, and C. $IU(w_1) = IU(A_1) + IU(A_2) + IU(C)$. In w_2, the objects of BITs realized are A_1 and B_2. $IU(w_2) = IU(A_1) + IU(B_2)$. $IU(w_1) > IU(w_2)$, assuming that all intrinsic utilities are positive and that $IU(C) > IU(B_2) - IU(A_2)$. In w_3, the objects of BITs realized are B_1, B_2, and C. $IU(w_3) = IU(B_1) + IU(B_2) + IU(C)$. In w_4, the objects of BITs realized are B_1 and A_2. $IU(w_4) = IU(B_1) + IU(A_2)$. $IU(w_3) > IU(w_4)$, assuming that $IU(C) > IU(A_2) - IU(B_2)$. These assumptions about BITs hold if $IU(A_1) = 6$, $IU(A_2) = 4$, $IU(B_1) = IU(B_2) = 3$, and $IU(C) = 2$. They are compatible with temporal utility's additivity for worlds. Reasons that support the order of worlds' rationality remove the objection to temporal utility's additivity.

Another example of temporal complementarity, aimed against additivity, arises from the measurement of health's temporal utility for a person. Suppose that constant health at level 2 each day of a ten-day period is preferable to health fluctuating between levels 1 and 3 from day to day during that period. Then, the temporal utility of health during a period is not the sum of the temporal utilities of health on each day in the period. The sum is twenty for both possibilities, but the first is preferable. A resolution of the problem observes that a rationale for the preferences, such as a basic intrinsic desire for stability of health, also defends temporal utility's additivity for the periods by introducing a new, relevant BIT.

Broome (1991: 229) presents the example in Table 4.3. T_1-T_4 are successive temporal intervals that constitute a world's duration. L_1-L_3 are lotteries, each of which has two possible outcomes, H and T, of equal probability. The lotteries may be coin tosses with heads and tails as possible outcomes. The table's cells present temporal utilities. The worlds of all three lotteries have an expected temporal utility of 100. A rational ideal agent may nonetheless prefer both L_1 and L_2 to L_3. Does summing temporal utilities during intervals go wrong?

Temporal-utility analysis handles the example. It assigns times of realization to relevant global features of a temporal sequence and counts these features when calculating the sequence's temporal utility. Most people have

Table 4.3 *Lotteries*

		T_1	T_2	T_3	T_4
L_1	H	100	-50	100	-50
	T	-50	100	-50	100
L_2	H	100	100	-50	-50
	T	-50	-50	100	100
L_3	H	100	100	100	100
	T	-50	-50	-50	-50

an aversion to risk and an attraction to security. A calculation of temporal utility attends to all BITs. It includes the value of security in appraisals of L_1 and L_2; these options guarantee a gain of 100 units. If the temporal utilities in the table's cells ignore the value of security, then its inclusion explains the rationality of preferences for L_1 and L_2. If the temporal utilities are comprehensive and register aversion to risk, then preferences for L_1 and L_2 are irrational. The preferences double count the value of security, once in the temporal utilities listed and a second time in discriminating among lotteries. Whether or not the temporal utilities are comprehensive, the preferences do not refute temporal-utility analysis.[7]

Temporal utility's separability overcomes McClennen's (1990: 206) objection that separability implies absence of commitment to past decisions and plans. Agents have reasons to honor past decisions and plans insofar as honoring them affects temporal utilities of futures. If an agent has a commitment to honoring a past decision, then honoring it raises temporal utility by realizing a BIT. Past choices affect the rationality of future choices by affecting their consequences.

Elga (2010) presents a case in which an agent has an opportunity to gain money for sure. For a hypothesis H, the agent can accept gamble A, which loses $10 if H and pays $15 if $\sim H$, and later she can accept gamble B, which pays $15 if H and loses $10 if $\sim H$. Accepting both gambles gains $5 for sure. If the agent rationally rejects A and later rejects B, then when she rejects B, she completes a sequence of acts that forgoes, without any compensation, the opportunity to gain money for sure. Rejecting B is irrational because it completes uncompensated rejection of an opportunity to gain money for

---

[7] Allais's paradox inspires the example. The paradox attacks the sure-thing principle, a principle of separability that belongs to expected-utility theory. The sure-thing principle survives the paradox if outcomes are comprehensive and include risk, as Weirich (1986) explains.

sure. Although rejecting B need not be irrational taken by itself, it is irrational when it follows rational rejection of A. Rejecting A is a type of sunk cost. Its effect on the status of rejecting B shows that sunk costs sometimes matter. The intuition against counting sunk costs assumes that the ranking of possible futures is independent of the past. Temporal-utility analysis handles the exceptional cases in which sunk costs matter because the past affects options' consequences. Given that an option's future includes the consequences it has because of its past, as the chapter's conventions of event location ensure, the temporal utilities of options' futures are separable from their common past.

Next, consider the money-pump objection to incomparable goods. It begins with an agent for whom x and y are incomparable goods and assumes that if $x+$ is a small improvement of x, then $x+$ and y are also incomparable. It also assumes that any trade among incomparables is rational, and that any trade yielding a gain is rational. A cycle of trades going from x to $x+$ to y to x is a money pump because the agent making the trades will pay to trade x for $x+$ but then ends up with x again without any gain, ready to pay once more for $x+$. All the trades in the cycle are rational, so the objection concludes that incomparability is the culprit. This objection loses force if an agent's past choices affect the utilities of outcomes of his current choices. Having rationally traded $x+$ for y, it is irrational to then trade y for x. Although the trade is rational taken in isolation, it is irrational when it completes the sequence of trades moving from $x+$ to x. Although an option's evaluation may ignore the past, the option's future includes the past's influence on it.

Temporal utility evaluates an option's comprehensive future. It conducts a comprehensive evaluation of an option's future. A comprehensive evaluation considers whatever the agent cares about. Temporal utility for an agent may accrue from the realization of any BIT, and a BIT's object may be almost any proposition. Because temporal utility evaluates an option's future comprehensively, any counterexample to a temporal-utility analysis must also use comprehensive evaluations. Examples using evaluations that omit considerations an agent cares about are not counterexamples. Some relevant considerations do not have natural, as opposed to conventional, times of realization. However, using conventional times of realization, temporal-utility analysis counts all relevant considerations and reliably compares the options in a decision problem.

Temporal-utility analysis's attention to all BITs does not give it a get-out-of-jail-free card. That is, it does not trivially gain immunity to counterexamples. Applications of temporal-utility analysis must specify an agent's BITs and cannot afterward introduce new BITs just to stave off

counterexamples. Of course, granting that temporal-utility analysis is correct, it has a type of immunity to counterexamples. Being a normative
principle, its truth is necessary. No possible case refutes it. Being a necessary
truth makes it immune to counterexamples without making it trivial.

Adding temporal-utility analysis to decision theory facilitates the theory's
resolution of decision problems by offering new ways of calculating options'
utilities and by simplifying utility comparisons of options. Temporal utilities of options' futures are easier to calculate than are options' comprehensive utilities because the temporal utilities put aside evaluation of the past.
The temporal utilities make precise traditional temporal evaluations of
options, as the next two sections show.

4.12 Hedonism

Temporal utility traditionally arises in evaluations of a person's life. The
main lines of these evaluations go back at least to Bentham ([1789] 1987).
The tradition takes a pleasure to have positive temporal utility and a pain to
have negative temporal utility. Because pleasures and pains are experiences,
they have periods of realization, namely, the periods during which they are
experienced. Hence, a lifetime's temporal utility divides according to times
during which temporal utility accrues. The temporal utility accruing during
a person's life is roughly her satisfaction during her life. Whereas a life's
comprehensive utility evaluates the life's entire world, a life's temporal
utility evaluates only the life itself. According to hedonism, a life's temporal
utility equals the sum of the temporal utilities of pleasures and pains during
the life, taking an experience's temporal utility to evaluate only the experience itself and to be roughly the degree of satisfaction it brings.

A calculation of temporal utility during a person's life first constructs a
lifetime temporal utility curve for the person. The curve represents the net
level of temporal utility the person experiences at each time during her life,
ignoring times of unconsciousness. For example, Figure 4.3 represents the
life of a person who experiences a higher and higher level of temporal utility
as her life goes forward. It takes the absence of pleasure and pain as temporal
utility's zero point, the start of consciousness as the start of the life's period,
and the end of consciousness as the end of the life's period.

A pleasure's temporal utility equals its intensity times its duration, and a
pain's temporal utility is the negative of its intensity times its duration.
Accordingly, the basic principle of calculation states that the amount of
temporal utility during a period of constant temporal-utility level equals
the level of temporal utility during the period times the duration of the

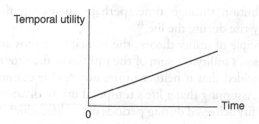

Figure 4.3 Temporal utility during a life

period. Application of integral calculus generalizes this principle for periods of variable temporal-utility level. The amount of temporal utility during a period is the limit of the sum of the temporal utilities of an approximating series of subperiods of constant temporal utility as they decrease toward zero width. This limit is the area under the temporal-utility curve for the period. The life's temporal utility, the amount of pleasure experienced discounted by the amount of pain experienced, is then the area under the whole curve.[8] In Figure 4.3, lifetime temporal utility is the area under the line. Calculation of lifetime temporal utility is an early form of temporal-utility analysis.

Temporal-utility analysis replaces the classical analysis's narrow, hedonistic conception of temporal utility with a broad conception, whereby nearly anything may affect temporal utility. Then, deriving the hedonistic calculations presented requires simplifying assumptions. A derivation may assume, for instance, that a life's temporal utility depends exclusively on the amounts of pleasure and pain experienced during the life. This simplifying assumption is plausible if only the desire for pleasure is intrinsic and basic, and if only the aversion to pain is intrinsic and basic.[9] In general, the simplifying assumptions have to justify analyzing a life's temporal utility into temporal utilities assigned to times during the life. They have to rule out the possibility that, for example, given an assignment of temporal utilities to times, a life's temporal utility depends partly on the temporal

[8] Beneath temporal utility's zero point, a curve's area is negative.
[9] The derivation of the analysis of lifetime temporal utility from temporal-utility analysis assumes that all and only psychologically basic pleasures are objects of basic intrinsic desires, that all and only psychologically basic pains are objects of basic intrinsic aversions, and that the intensities of these basic intrinsic attitudes go by amounts of the psychologically basic pleasures and pains. It also assumes that temporal utility attaches to particular pleasures that propositions represent. Conjunctions of such propositions represent combinations of basic pleasures. Pains have a similar propositional representation.

utilities' distribution through time, perhaps increasing if the temporal utilities steadily rise during the life.[10]

A basic principle of utility theory, the principle of pros and cons, states that a proposition's utility is a sum of the utilities of the pros and cons of its realization, provided that it neither omits nor double counts any relevant consideration. Assuming that a life's temporal utility divides into amounts of temporal utility achieved during periods of the life, which are themselves broken down into temporal utilities assigned to times during the periods, the principle of pros and cons supports the traditional method of analyzing lifetime temporal utility. The total temporal utility during a life stems from temporal utilities during periods where the balance of pleasure and pain is constant. Net pleasure during a period is a pro, whereas net pain is a con. The division of pros and cons according to periods neither omits nor double counts any relevant consideration because by assumption temporal utility is assigned to times, which are neither omitted nor double counted. The product of the intensity and the duration of the net pleasure or pain experienced during a period weights the pro or con the period represents. After the analysis weights temporal pros and cons, it adds them to obtain the total temporal utility accruing during the life.

Temporal-utility analysis generalizes the analysis of lifetime temporal utility. It applies to an arbitrary proposition, not just one expressing a life or a period of a life. The analysis of lifetime temporal utility follows from temporal-utility analysis's application to a proposition expressing a life. The duration of the life is the proposition's interval of realization. The proposition that one leads the life during a subinterval has the subinterval as the interval of realization. Given the assumption that BITs attend to pleasure and pain exclusively, a curve indicating the balance of pleasure over pain at each time during the life provides the temporal utilities assigned to times, and the area under the curve is the life's temporal utility.

Temporal-utility analysis discharges the analysis of lifetime temporal utility's restrictive assumptions about BITs. As a result, events partly outside a person's lifetime may affect the temporal utility of the person's life. For example, a baron may have a basic intrinsic desire that his ancestor's manor be restored. The baron's family may have started the restoration before his birth. Completion of the restoration during his lifetime boosts his lifetime temporal utility, although the restoration does not occur entirely during his

---

[10] Velleman (2000) argues that because the distribution of well-being during a life affects lifetime well-being, lifetime well-being is not a sum of well-being at times during the life.

lifetime. Temporal-utility analysis has a broader perspective than does the analysis of lifetime temporal utility.

4.13   Mean-risk analysis

Temporal-utility analysis enriches utility theory many ways. As an illustration, this section shows how it yields a version of a common investment tool, mean-risk–utility analysis. This form of analysis evaluates investments according to average or mean return, and also risk. It takes risk as an interaction effect of the various chances for gain or loss that an investment generates. Although risk in the ordinary sense includes the chances themselves, risk as a component of mean-risk–utility analysis excludes them because mean return covers them.

Mean-risk–utility analysis rests on assumptions about basic intrinsic attitudes and risk. Suppose that a person has a basic intrinsic aversion to risk (in the subjective sense that makes risks depend on subjective probabilities). He may separate evaluation of realization of that aversion from evaluation of realization of other basic intrinsic attitudes (BITs). He may also temporally separate the evaluations. When he adopts a risky option, the risk accrues immediately. Hence, the aversion's time of realization is the time of the option's adoption.[11] Suppose that the aversion to risk is the only BIT realized then. In this case, the temporal utility of the option's future divides into the temporal utility of realization of the aversion to risk and the temporal utility of realization of other BITs. Let $R[o]$ express realization of the risk involved in an option o, let $S[o]$ state that o's sequel obtains, where o's sequel consists of the events following o. If f_i specifies a possible future of o, then $f_i - o$ specifies o's sequel in that possible future, that is, o's possible future f_i excluding o. An option's sequel excludes the option, and its future is the option together with its sequel. Given complete information, a temporal-utility analysis of o's future with respect to a twofold partition of its duration yields the following equation: $TU(F[o]) = TU(R[o]) + TU(S[o])$. $R[o]$'s interval of realization is the moment of o's adoption, and $S[o]$'s interval of realization is the interval following o's adoption. Applying expected-utility analysis to the latter, to

---

[11] In a typical case of a new product's introduction, the risk lasts during a lengthy period, for example, the period during which the new product is designed, manufactured, and marketed. This happens because the product's introduction occurs during an interval and not just at a moment. Also, the time of realization of a risk when one buys stocks seems to be not the moment of the decision to purchase them, but the entire interval during which one owns the stocks. The explanation is that owning stocks is a new gamble from moment to moment. Each moment one owns them, one runs a new risk of depreciation that could have been avoided by selling.

accommodate incomplete information, transforms it into the expected temporal utility of a lottery over o's possible sequels. Substituting $F'[o]$, the proposition that $F[o]$ obtains, to accommodate incomplete information about o's future, the equation becomes $TU(F'[o]) = TU(R[o]) + \sum_i P(f_i$ given $o) TU(f_i - o)$, where f_i ranges over possible futures of o. This equation expresses the temporal utility of o's future in terms of the temporal utility of o's risk and a mean temporal utility of o's possible sequels. It is a form of mean-risk–utility analysis in which the mean is a probability-weighted average of temporal utilities of possible sequels.[12]

Mean-risk–utility analysis, formulated using temporal utility, has temporal-utility analysis's support. It illustrates the type of temporal separation that justifies simplification of choices. Because of such separation, deliberations may hold fixed one temporal interval and focus on events in other temporal intervals.

This chapter introduced temporal utility and then, using conventions of temporal location, showed how to obtain the temporal utility of a period from the temporal utilities of subperiods. This temporal-utility analysis justifies evaluating the options in a decision problem according to their futures. It puts a commonplace method of simplifying choices on firm footing.

4.14 Appendix: Consistency of utility analyses

Temporal-utility analysis uses a temporal type of utility, differing from the comprehensive type of utility decision theory usually employs. However, the temporal utility of an option's outcome equals the comprehensive utility of the option's outcome, that is, the option's comprehensive utility. This appendix shows that an option's comprehensive utility, as computed by a temporal-utility analysis of its outcome, agrees with a canonical utility analysis of the option's comprehensive utility. An analogous result holds for conditional utilities.

Consider an option o and its outcome $O[o]$. Because $O[o]$'s interval in a world where o is realized is the world's duration, applying Section 4.8's general form of temporal-utility analysis,

$$U(o) = TU(O[o]) = \sum_i P(w_i \text{ given } o) \sum_j TU(w_{ij})$$

---

[12] Under this section's assumptions about BITs and their times of realization, and given the reduction of comprehensive-utility comparisons of options to temporal-utility comparisons of their futures, the new form of mean-risk analysis is equivalent to the form Weirich (2001: sec. 5.4) presents. A demonstration shows that both forms of mean-risk analysis are reducible to a canonical utility analysis, following Section 4.14's pattern for the reduction of temporal-utility analysis to intrinsic-utility analysis.

where w_{ij} ranges over propositions characterizing the elements of a partition of w_i's duration. By the definition of temporal utility, the canonical way of obtaining input for a temporal-utility analysis, $TU(w_{ij})$ is the sum of the intrinsic utilities of the objects of the BITs realized at times in w_{ij}'s interval. That is, $TU(w_{ij}) = \sum_k IU(\text{BIT}_{ijk})$, where BIT_{ijk} ranges over the objects of the BITs realized at times in w_{ij}'s interval. Hence, $TU(O[o]) = \sum_i P(w_i$ given $o)\sum_j \sum_k IU(\text{BIT}_{ijk})$. Because each BIT realized in a world is assigned to exactly one element of a partition of the world's duration, by algebra,

$$TU(O[o]) = \sum_i P(w_i \text{ given } o)\sum_j IU(\text{BIT}_{ij})$$

where BIT_{ij} ranges over the objects of BITs realized in w_i. Section 3.8's canonical calculation of $U(o)$ using BITs yields the same sum, so the temporal-utility analysis of $U(o)$ using $TU(O[o])$ agrees with its canonical analysis. Temporal-utility analysis is consistent with all forms of utility analysis that reduce to the canonical form of analysis.

CHAPTER 5

Spatiotemporal utility

A homeowner drafting plans for a remodeling project imagines removing a wall to enlarge a room. Will removing the wall make the roof sag? Effects of the wall's elimination extend forward in time and also spatially, possibly from the room to the roof. The wall's elimination may propagate a chain of events with distant effects as time progresses. Its removal may cause the roof to sag and then leak, generating repairs that gather supplies from surrounding stores and pay workers who send income taxes to the capital. To evaluate a decision to remove the wall, the homeowner may evaluate just events in the decision's region of influence. This region includes every relevant event.

Because an option's effects spread through space as well as time, they occupy a spatiotemporal region. The option's evaluation may focus on the region in which the option's effects occur. Instead of appraising an option's future, as an option's temporal evaluation does, an option's spatiotemporal evaluation appraises events in the option's region of influence. This is a smaller portion of the option's world than its future. Evaluating it simplifies deliberations.

This chapter explicates and supports a common spatiotemporal method of evaluating an option. The method spreads an option's pros and cons through space and time and then evaluates the pros and cons in the option's region of influence. Its explication introduces spatiotemporal utility and spatiotemporal-utility analysis. Spatiotemporal utility reduces the scope of an option's evaluation to events in the option's region of influence, and spatiotemporal-utility analysis divides events in an option's world into those inside and outside an option's region of influence. It simplifies by dividing an option's world into a part to evaluate and a part to ignore.

As Chapter 4 assigns events to times, this chapter assigns events to space-time points, using conventions to supplement natural assignments. The assignments ensure that spatiotemporal-utility analysis neither omits nor double counts relevant considerations. Given the chapter's idealizations, the method of analysis justifies evaluating the options in an agent's decision

problem according to events in the options' regions of influence. Although
an option's region of influence includes remote events, if the agent does not
care about them, the option's evaluation omits them. If they are not the
option's consequences, Chapter 6 puts them aside. Spatiotemporal utility
yields an efficient but not a maximally efficient general method of evaluating
options that processes all relevant considerations.

5.1   An option's region of influence

Spatiotemporal utility continues the simplification of deliberation that
temporal utility started. Temporal utility pares away events prior to an
option's realization. It evaluates an option using the four-dimensional world
branch that would obtain if the option were realized. Spatiotemporal utility
pares away even more events than does temporal utility. It evaluates an
option using only the part of its world branch in the option's region of
influence.

Temporal utility's introduction began with an assignment of events to
intervals, including the duration of an option's world branch, or the
option's future. Spatiotemporal utility's introduction similarly starts with
an assignment of events to spatiotemporal regions, including an option's
region of influence. This section defines a proposition's region of influence
and characterizes the events it contains.

A world has a space-time structure. Different worlds may have different
space-time structures. Relativity theory makes simultaneity observer dependent, and using the agent in a decision problem as the observer yields a
suitable framework for the temporal relations that spatiotemporal-utility
analysis uses. A spatiotemporal-utility analysis of an option's world partitions
the BITs the world realizes so that the analysis counts each realization of a
BIT but counts none twice. A world's space-time structure and the location
in it of realizations of BITs supply a means of partitioning realizations of
BITs. Illustrations assume that space-time has a structure that permits a four-
dimensional representation using a Minkowski diagram. Such a diagram aids
visualization of the partition. Spatiotemporal-utility analysis accommodates
a space-time structure without a Minkowski representation (because space
curves or time is cyclical), assuming only locations for realizations of BITs at
positions in the structure and a partition of the structure.

Spatiotemporal utility evaluates a proposition by evaluating events in the
proposition's region of influence. This is the spatiotemporal region of the
proposition's world that begins with the proposition's realization and
stretches as far in space-time as effects of the proposition's realization may

extend. Given that the speed of light limits causation's reach, the region is the future-directed light cone, or more briefly, the light cone of the proposition's realization. An option's *region of influence* consists of the space-time points where the option may influence events. A proposition expressing a spatiotemporally extended event has an extended *region of realization* as well as an extended region of influence. Although an option has an extended region of influence, by idealization, a decision's and so an option's region of realization is a space-time point.

Given a proposition's realization, this chapter calls the events in the proposition's region of influence the proposition's *result*. A proposition p's result, $R[p]$, stands for another proposition that specifies the occurrence of the events that would occur in p's region of influence if p were realized, including p's realization. Efficient deliberations limit the specification to significant events, that is, events an agent cares about and so occur in p's world as Chapter 2 characterizes it. An option's result plays in spatiotemporal-utility analysis the role that the option's future plays in temporal-utility analysis.

Given complete information, an agent knows $R[p]$. When comprehensive utility applies to $R[p]$, it evaluates the outcome of $R[p]$, namely, all that would happen if $R[p]$ were to obtain, including events outside p's region. Because p and $R[p]$ have the same outcome, p's comprehensive utility equals $R[p]$'s comprehensive utility. That is, $U(p) = U(R[p])$. If the spatiotemporal utility of p, $SU(p)$, were equal to $U(R[p])$, $U(p)$ would equal $SU(p)$ so that spatiotemporal utility would not diminish comprehensive utility's scope. However, $SU(p)$ evaluates just the events in p's spatiotemporal region of realization, and not the events in p's region of influence or the events in p's world. To evaluate a proposition o that expresses an option, deliberations use $SU(R[o])$. This type of utility evaluates all events in $R[o]$'s region of realization, that is, o's region of influence.

A proposition's spatiotemporal utility assumes a separation of events in the proposition's region of influence from the other events that would obtain if the proposition were realized. Different methods of separation may suit different applications of spatiotemporal utility. This section settles the separation for a proposition expressing an option in a decision problem, leaving other cases open. The separation for an option assigns to an option's region of influence any event that is partly in the option's region of influence; if the option may influence part of an event, it may influence the event. Spatiotemporal utility's evaluation of the option's region of influence considers every overlapping event to ensure counting every event relevant to the option's evaluation.

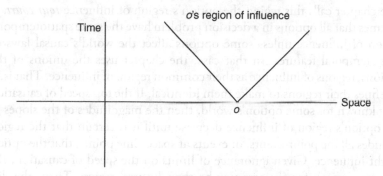

Figure 5.1 An option's region of influence

To obtain an option's region of influence in a world that realizes the option, begin with the option's future in the world. A time line representing it thickens to represent the future's spatial dimensions. Excising space-time points beyond the option's influence yields the option's region of influence. Figure 5.1 shows a two-dimensional projection of an option *o*'s four-dimensional region of influence.

The points in the diagram are space-time coordinates. The horizontal line and the area above it represent the region of the option's future. The vee opening to the top and its interior represent the option's region of influence, and the vertex of the vee represents the space-time point of the option's realization. The slope of the right boundary of the option's region of influence depends on causation's top speed in the world realizing the option. If that speed equals light's speed, then its slope equals the inverse of the speed of light. The left boundary's slope has the same magnitude, so that the region contains all the space-time points where the option may influence events by light emission. The option's region of influence is thus the option's future-directed light cone. Assuming special relativity theory, the region of influence of any point-event is the event's future-directed light cone, and the region that influences any point-event is its past-directed light cone.

The chapter's illustrations assume that an option's region of influence is its future-directed light cone. The assumption rests on the empirical law that the speed of light limits causation's speed. However, spatiotemporal-utility analysis works for an agent given any region of influence such that the agent is certain that options' effects occur in that region. The analysis's form does not depend on empirical assumptions about causation's speed.

In a decision problem, an option's evaluation targets the option's region of influence in the world that would be realized if the option were realized.

The chapter calls this region the option's region of influence *tout court*. It assumes that all options in a decision problem have the same spatiotemporal region of influence unless some options affect the world's causal laws or spatiotemporal features. In that case, the chapter uses the unions of the options' regions of influence as their common region of influence. That is, it redefines their regions to make them identical. If the top speed of causation is unknown for some option's world, then the magnitudes of the slopes of the option's region of influence decrease until it is certain that the region includes all the point-events, or events at space-time points, that the option might influence. Given ignorance of limits on the speed of causation, the common region for options may be their futures' region. Then, the distinction between spatiotemporal and temporal utility disappears. Given backward causation, the options' region of influence may be larger than their futures' region.

Although spatiotemporal-utility analysis does not assume that the speed of light limits the speed of an act's effects, the analysis's efficiency depends on knowledge of limits for causation's speed. Given knowledge of a limit, spatiotemporal-utility analysis is more efficient than temporal-utility analysis. This comparative advantage also obtains in an agent's decision problem if the agent knows that all relevant effects propagate no faster than some finite speed. In contrast, given complete ignorance of causation's speed, spatiotemporal-utility analysis reduces to temporal-utility analysis and so loses its comparative advantage. The chapter's normative model gives agents knowledge of causation's speed.

Spatiotemporal evaluation of an option uses the option's region of influence in the option's world, but for completeness, the characterization of regions of influence treats any proposition in any world that realizes the proposition. A proposition's region of influence in a world is the union of the regions of influence of the point-events in the region of the proposition's realization in the world. A possible world's region of influence is the world's region of realization. Convention may settle a proposition's region of realization if intuition leaves its boundary vague and thereby settle the proposition's region of influence.

In an option's world, the events in the option's region of influence include, first, the point-events in the option's region of influence. These are all the point-events that the option may influence as well as those that the option in fact influences. None precedes the option. Among point-events, transitivity governs the relation of being-in-the-region-of-influence-of. Take point-events x, y, and z. If x is in the region of influence of y, and y is the region of influence of z, then x is in the region of influence of z.

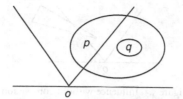

Figure 5.2 Transitivity's failure

Next, the option's region of influence includes extended events with point-events in the option's region of influence; any extended event with a point-event in the option's region of influence is in the option's region of influence. Chapter 4 similarly puts an event into an option's future if any part of it is in the option's future. In general, an option's evaluation assigns to an option's zone of evaluation any event that overlaps the zone.

According to this specification of events in an option's region of influence, an option *o* may have in its region of influence *p* because part of *p*'s region of realization is in *o*'s region of influence. Also, *p* may have *q* in its region of influence because *q*'s entire region of realization is in *p*'s region of realization, and so also in *p*'s region of influence. However, if none of *q*'s region of realization is in the option's region of influence, then the option does not have *q* in its region of influence. Hence, region-events prevent transitivity for the relation of being-in-the-region-of-influence-of. As Figure 5.2 shows, the event *o*, an option whose region of influence is its light cone, has *p* in its region of influence, and the event *p* has *q* in its region of influence, but the event *o* does not have *q* in its region of influence. This failure of transitivity is not counterintuitive because it involves extended events rather than point-events.

The last moment of an event's realization may fall outside an option's region of influence, although some moments of its realization fall inside the option's region of influence. Then, the event is in the option's region of influence even if its moment of realization (the last moment of its realization) is not in that region. Figure 5.3 illustrates this possibility. The event *e* that the triangle represents (it could be a BIT's realization) is partly in the option *o*'s light cone, but its time of realization, which the triangle's top vertex represents, is outside the option's light cone.

This section's account of the events in an option's region of influence grounds the next section's definition of an option's spatiotemporal utility. For an option and each world that might be the option's world, an ideal agent knows the assignment of events to the option's region of influence,

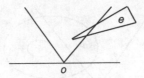

Figure 5.3 Influence without completion

granting the conventions for the assignment. Consequently, given the next section's definition of an option's spatiotemporal utility, an ideal agent knows the option's spatiotemporal utility.

5.2 Spatiotemporal utility defined

The intrinsic utility of a basic intrinsic attitude's realization is the fundamental particle to which Chapter 3 reduces comprehensive utility, given complete information. It provides a means of defining spatiotemporal utility. This section presents first a definition of spatiotemporal utility for cases of complete information and then generalizes the definition to cover cases of incomplete information.

A proposition p's *spatiotemporal utility*, $SU(p)$, depends on the intrinsic utilities of events in p's spatiotemporal region of realization, supposing its realization. $SU(p)$ evaluates p according to the basic intrinsic attitudes (BITs) realized in p's region of realization. These BITs ground the intrinsic utilities of other events in p's region of realization. Given complete information, $SU(p) = \sum_i IU(\text{BIT}_i)$, where BIT_i ranges over the objects of BITs whose realizations occur in p's region of realization. That is, the spatiotemporal utility of p is the sum of the intrinsic utilities of the objects of the BITs that would be realized in p's region of realization if p were realized. Allowing for incomplete information, $SU(p)$ equals the expected value of that sum, as the remainder of this section explains.

That SU's application to a proposition depends on the proposition's region of realization does not alter application of other propositional functions, such as an agent's probability function. Probability's application to a proposition depends just on the probabilities of worlds in which the proposition holds. Chapter 2's interpretation of probability as rational degree of belief is constant.

Spatiotemporal utility recognizes that a proposition's region of realization may vary with the worlds realizing the proposition. A disjunction, for example, may have different regions of realization in a world where its first disjunct's realization makes it true and in a world where its second disjunct's realization makes it true. A proposition p's result $R[p]$ is a proposition fully specifying relevant events in p's region of influence in p's world. Even given idealizations about agents' cognitive powers, $R[p]$ may be unknown, so its comprehensive utility $U(R[p])$ may also be unknown. Let $R'[p]$ be the proposition that $R[p]$ obtains. An ideal agent knows $U(R'[p])$ even if she does not know $U(R[p])$. $U(R'[p])$ is the expected value of $U(R[p])$. Given full information, $U(R[p])$ equals $U(R'[p])$, but given incomplete information, the utilities may differ.

As Section 5.1 notes, $R[p]$'s region of realization is the same as p's region of influence. Because the BITs whose realizations $R[p]$ entails are the same as those whose realizations obtain in p's region of influence, $IU(R[p])$ evaluates events in p's region of influence. Assuming complete information, $SU(R[p]) = IU(R[p])$. Also, $IU(R[p]) = \sum_i IU(\text{BIT}_i)$, where BIT_i ranges over the objects of BITs whose realizations occur in p's region of influence. So $SU(R[p]) = IU(R[p]) = \sum_i IU(\text{BIT}_i)$, where BIT_i has the same range.

Suppose that a proposition o expresses an option in a decision problem. The spatiotemporal utility of the option o, assuming complete information, is the sum of the intrinsic utilities of the objects of BITs whose realizations are assigned to the option's region of realization. In other words, $SU(o) = \sum_i IU(\text{BIT}_i)$, where BIT_i ranges over the objects of BITs whose realizations in o's world occur in o's region of realization. This utility ignores o's effects. A more comprehensive evaluation of o uses the spatiotemporal utility of its region of influence rather than its region of realization. The option's result $R[o]$ specifies the events in o's region of influence. So the evaluation assesses $R[o]$. It uses $SU(R[o])$.

The definition of spatiotemporal utility ensures a precise value for a proposition's spatiotemporal utility only given a specification of the events in the proposition's world that are in the proposition's region of realization. Section 5.1 specifies these events for propositions expressing the regions of influence of the options in a decision problem but leaves open the specification for other propositions.

How does the definition of spatiotemporal utility apply to an option's result given the possibility of incomplete information, in particular, incomplete information about the effects of an option's realization? Given incomplete information, it targets $SU(R'[o])$ instead of $SU(R[o])$, its target given complete information. $SU(R'[o])$ equals the expected value of $SU(R[o])$. The following paragraphs explain this equality.

Given incomplete information, a proposition's comprehensive utility evaluates a lottery over the possible worlds that realize the proposition. The comprehensive utilities of the possible worlds derive from the BITs the worlds realize. The proposition's comprehensive utility is thus the expected intrinsic utility of the realizations of BITs in worlds that realize the proposition.

Spatiotemporal utility uses a similar method of evaluation. For an ideal agent who assigns and knows the probabilities and utilities that an expected-utility analysis presumes, a proposition's spatiotemporal utility is the expected intrinsic utility of events in the proposition's region of realization. It is calculated using intrinsic utilities of possible full descriptions of the events in the proposition's region of realization and the descriptions' subjective probabilities given the proposition. Unlike a proposition's comprehensive utility, a proposition's spatiotemporal utility ignores events not in the proposition's region of realization. Its spatiotemporal utility, although also an evaluation of a lottery over intrinsic utilities of objects of sets of BITs realized in worlds where the proposition is true, takes the set for a proposition and world to include only BITs realized in the proposition's region of realization (according to conventions for settling the proposition's region of realization and the location of BITs' realizations).

Thus, the general definition of a proposition p's spatiotemporal utility is

$$SU(p) = \sum{}_i P(r_i \text{ given } p)IU(r_i)$$

where r_i ranges over full descriptions of p's possible realizations, that is, full descriptions of the events constituting p's possible realizations. Because r_i is a full specification of p's realization, incomplete information does not affect its evaluation. $SU(r_i) = IU(r_i)$. So, the general definition immediately implies that $SU(p) = \sum{}_i P(r_i \text{ given } p)SU(r_i)$. In other words, a proposition's spatiotemporal utility is its expected spatiotemporal utility taken with respect to a fine-grained partition of its possible realizations.

The definition of spatiotemporal utility uses specifications of regions of realization. The definition's immediate implications use BITs realized in the regions specified. The conjunction of objects of BITs a region realizes may represent the region. The intrinsic utility of a proposition's possible realization derives from the BITs whose realizations occur in the proposition's region of realization. A possible realization's intrinsic utility is the sum of the intrinsic utilities of the objects of BITs realized in the proposition's region of realization according to the possible realization. So $SU(p)$ arises from fundamental intrinsic utilities:

$$SU(p) = \sum{}_i P(r_i \text{ given } p)\sum{}_j IU(\text{BIT}_{ij})$$

where r_i ranges over possible full specifications of p's region of realization, and BIT_{ij} ranges over BITs whose realizations r_i entails.

The foregoing presentation of a proposition's spatiotemporal utility takes it as an estimate of the intrinsic utility of the BITs realized in the proposition's region of realization. The formula for the proposition's spatiotemporal utility advances a pros and cons evaluation of the proposition's realization. According to the formula, a proposition's spatiotemporal utility stems from the spatiotemporal utilities of certain chances (resting on subjective probabilities) that accompany its realization, namely, chances for possible sets of events in its region of realization. The chances are sure accompaniments of the proposition's realization, despite uncertainty about other events accompanying its realization. A chance's spatiotemporal utility equals the product of the probability of and the intrinsic utility of a possible specification of events in the proposition's region of realization. The sum of the spatiotemporal utilities of the chances equals the proposition's spatiotemporal utility.

For an option o, the definition of its region of influence's spatiotemporal utility $SU(R[o])$ uses a specification of events in the option's region of influence r_i. Given incomplete information, the definition of $SU(R'[o])$ uses all r_i that are possible denotations of $R[o]$. $SU(R'[o])$ equals the expected value of $SU(R[o])$. It evaluates an option's region of influence given incomplete information by evaluating the option's possible results, that is, the events that might occur in the option's region of influence.

Conditional spatiotemporal utility has a straightforward introduction. Given complete information, $SU(p$ given $q)$ is the intrinsic utility of a full specification of p's region of realization given q. Using basic intrinsic attitudes,

$$SU(p \text{ given } q) = \sum_i IU(\text{BIT}_i)$$

where BIT_i ranges over the objects of BITs whose realizations occur in p's region of realization given q. Allowing for incomplete information, $SU(p$ given $q)$ is a probability-weighted average of the spatiotemporal utilities of possible full descriptions of p's region of realization given q. $SU(p$ given $q) = \sum_i P_q(r_i$ given $p)IU(r_i)$, where r_i ranges over possible full specifications of p's region of realization given q, and P_q yields probabilities conditional on q. $P_q(r_i$ given $p) = P(r_i$ given $(p \ \& \ q))$ in typical cases of supposition of p under supposition of q. Using basic intrinsic attitudes,

$$SU(p \text{ given } q) = \sum_i P_q(r_i \text{ given } p)\sum_j IU(\text{BIT}_{ij})$$

where r_i ranges over possible full specifications of p's region of realization given q, and BIT_{ij} ranges over the objects of BITs realized in p's region of realization according to r_i.

Comprehensive utility and spatiotemporal utility evaluate an option using different intrinsic utilities. The definition of spatiotemporal utility shows that the spatiotemporal utility of an option o's result $R[o]$ evaluates just events in o's region of influence instead of the whole outcome of o's realization. $W[o]$ is a full specification of o's world, and $R[o]$ is a full specification of events in o's region of influence. The intrinsic utilities $IU(W[o])$ and $IU(R[o])$ are distinct because the BITs whose realizations $W[p]$ entails differ from those whose realizations $R[o]$ entails. Given complete information, $U(o) = IU(W[o])$, and $SU(R[o]) = IU(R[o])$. Given incomplete information, $U(o)$ and $SU(R'[o])$ equal the expected values of the corresponding intrinsic utilities.

5.3 Analysis of spatiotemporal utilities

Spatiotemporal-utility analysis breaks the spatiotemporal utility of a proposition into the spatiotemporal utilities of other propositions. Section 5.2's formula, $SU(p) = \sum_i P(r_i \text{ given } p)SU(r_i)$, where r_i ranges over full descriptions of p's possible realizations, supplies a type of spatiotemporal-utility analysis. This section introduces a type of spatiotemporal-utility analysis that divides a proposition's spatiotemporal utility according to a partition of the proposition's spatiotemporal region.

A farmer may care about the north forty acres and about the south forty acres. A federal official may care about each state in the union. Spatiotemporal-utility analysis offers a way of evaluating disjoint spatiotemporal regions and combining their evaluations to obtain an evaluation of the regions' combination. In a spatiotemporal diagram such as Figure 5.1, a region may be Missouri's shape projected into the future. Or, it may be a space-time worm comprising the times and places of a drugstore manager's activities throughout a business year. Adjacent cubes, or a tangle of space-time worms, projecting from the horizontal axis for the present may represent a combination of regions.

Spatiotemporal-utility analysis according to a partition of a spatiotemporal region explicates a traditional form of analysis that divides an event's region of influence according to times and places. Consider the spatiotemporal utility of a hurricane. Its computation may evaluate events in the hurricane's path on a sequence of days in contiguous areas. Similarly, computation of the spatiotemporal utility of a donation to charity may evaluate events at times in the lives of people in places where the money was spent. A division of events according to times and places prevents omitting or double counting relevant events.

Because intuition may not assign events to precise spatiotemporal regions, some applications of spatiotemporal-utility analysis adopt a conventional assignment. When intuitions are insufficient, conventions assign an event to an element of a region's partition. For a BIT the region realizes, conventions assign its realization to a subregion if the BIT's realization does not naturally fall into exactly one subregion. An analysis of a proposition's spatiotemporal utility according to a partition of its region of realization ensures that the analysis neither omits nor double counts BITs realized in the region.

For a proposition p, a spatiotemporal-utility analysis of $SU(p)$ separates events in p's region of realization using a partition of the region. The proposition p's spatiotemporal utility is the sum of the spatiotemporal utilities of events in the elements of the partition. According to the analysis, each BIT whose realization occurs in p's region of realization occurs in just one element of its partition. Thus, for a proposition p,

$$SU(p) = \sum_j SU(p_j)$$

where p_j ranges over propositions specifying events in the elements of a partition of p's region of realization. Suppose that w_i is p's world. Let r_{ij} range over full specifications of the events in p's region of realization in w_i assigned to the jth element of a partition of the region. $IU(r_{ij})$ is the sum of the intrinsic utilities of objects of BITs realized according to r_{ij}. Because $SU(p_j) = IU(r_{ij})$, it follows that $SU(p) = \sum_j IU(r_{ij})$.

A partition of p's region of realization frames a division of the BITs assigned to the region. Each element of the region's partition may, for example, contain just the events in a place in p's region of realization or just a person's activities in p's region of realization. The space-time worms for distinct places and for distinct people do not intersect, even if the places or people causally interact, and so they may be elements of a partition of a proposition's region of realization.

The formula for spatiotemporal-utility analysis accommodates incomplete information because a proposition's expected spatiotemporal utility equals its spatiotemporal utility given the chapter's idealizations about agents. Supplementary formulas for a proposition's spatiotemporal utility make this explicit.

First, applying expected-utility analysis to a proposition p's spatiotemporal utility yields the formula

$$SU(p) = \sum_i P(s_i \text{ given } p)SU(p \text{ given } s_i)$$

where s_i ranges over possible states of the world. In some cases, the states are possible full specifications of p's region of realization.

Next, applying expected-utility analysis and spatiotemporal-utility analysis simultaneously yields a hybrid form of utility analysis:

$$SU(p) = \sum_i P(s_i \text{ given } p)\sum_j SU(p_j \text{ given } s_i)$$

where s_i ranges over possible states of the world, and p_j ranges over specifications of the events in the elements of a partition of p's region of realization given s_i. There are multiple ways of calculating $SU(p)$ using partitions of states of the world and partitions of p's region of realization.

All the book's forms of utility analysis apply to a proposition's spatiotemporal utility. For instance, an analysis may use expected-utility analysis as shown earlier to compute a proposition's spatiotemporal utility in cases in which the state of the world is uncertain. In general, an analysis may use any form of utility analysis that works by separation of BITs on any type of utility that evaluates according to BITs. A proposition's spatiotemporal utility equals the expected value of the sum of the intrinsic utilities of the objects of BITs realized in the proposition's region of realization, in cases of complete and incomplete information alike, and given conditions. An analysis may apply together all forms of utility analysis resting on BITs – in particular, combinations of intrinsic-, expected-, and temporal-utility analyses. A reduction of each form of utility analysis to the same canonical form of utility analysis using BITs ensures consistency. The chapter's appendix supports the consistency of these methods of analysis applied to spatiotemporal utility.

5.4   Location of events

Spatiotemporal utility evaluates a proposition by evaluating events in the proposition's spatiotemporal region of realization. It assumes that propositions have spatiotemporal regions of realization and that events belong to spatiotemporal regions. Spatiotemporal-utility analysis sums some spatiotemporal utilities to obtain others. A spatiotemporal-utility analysis presumes an assignment of events to spatiotemporal regions according to which the analysis does not omit or double count relevant events. Section 5.1 specifies an assignment of events to an option's region of influence that suits a spatiotemporal-utility analysis of an option's world using the option's region of influence and its complement. This section briefly presents an alternative assignment of events to regions that suits other forms of spatiotemporal-utility analysis.

Propositions express events, and the location of events goes by the location of realizations of propositions that express the events. Assigning propositions' realizations to spatiotemporal regions raises complex issues. For spatiotemporal-utility analyses, an assignment that does not conflict with intuition is satisfactory. Congruence with intuition is unrealistic because intuition does not assign each proposition to a precise spatiotemporal region in every world that realizes the proposition. Spatiotemporal-utility analyses use conventions to supplement intuition and to assist specification of a proposition's region of realization so that each proposition a world realizes has a precise region of realization in the world. If intuition leaves vague a proposition's region of realization in a world, convention makes it definite.

Temporal-utility analysis uses an event's time of realization to assign the event to a temporal interval. Some spatiotemporal-utility analyses assign an event to a spatiotemporal region according to the event's point of realization. Although an extended event's location is a region, an analysis may assign the event's completion to a point, the event's realization point, to settle the event's assignment to a region. A proposition's realization point in a world belongs to the proposition's region of realization in the world. It may vary with the possible world in which the proposition holds. The realization point for a proposition expressing a point-event is the event's space-time point in the world realizing the proposition. The realization point for a proposition expressing a region-event is the last space-time point in its region of realization in a world; or if several last points exist, an arbitrary last point; or if no last point exists, an arbitrary point in its region (in a decision problem, a point in the options' region of influence if the region-event extends into that region). Accordingly, for two propositions a world realizes, the first proposition's realization is in the second proposition's region of realization if and only if the first proposition's point of realization is in the second proposition's region of realization.[1]

Some applications of spatiotemporal-utility analysis assign events to regions without using realization points. For comparison of options in a decision problem, an analysis may settle whether a proposition's realization belongs in an option's region of influence. Section 5.1 puts a proposition's realization in an option's region of influence if and only if the proposition's

---

[1] In decision problems, worlds trim irrelevant detail. If in a decision problem, two untrimmed worlds differ about a proposition's realization point, and the difference affects assignment of a relevant event to a region of a spatiotemporal-utility analysis, then the difference counts as relevant, and trimmed worlds retain identification of the proposition's realization point.

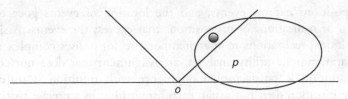

Figure 5.4 A proposition's point of realization

region of realization overlaps the option's region of influence. A conventional assignment of events to space-time points, different from the previous paragraph's, may duplicate its assignment of events to the option's region of influence.

Suppose that for a proposition p expressing an event, the proposition's region of realization in a world overlaps the region of influence for an option o realized in the world. The new conventional assignment makes p's realization point a coordinate in o's region of influence. Realization of p belongs to o's region of influence if and only if its realization point is in o's region. So p's realization counts as an event in o's region of influence. In other words, in a world where p and o are realized and a point in p's region of realization is in o's region of influence, the new conventional assignment puts p's realization point in o's region of influence. Figure 5.4 illustrates the assignment. It does not use the completion of p's realization, the last point in p's region of realization, as p's point of realization.

According to the new conventional assignment, any event with parts in an option's region of influence is in the option's region of influence. If a proposition's realization is part of the outcomes of two options in a decision problem and has the same region of realization given each option, then it is in one option's region of influence if and only if it is in the other option's region of influence, because the options' regions of influence are the same. The assignment puts extended events straddling the moment and place of a decision problem in each option's region of influence.

An apt assignment of events to regions depends on spatiotemporal-utility analysis's application – its object, elements, and purpose. Consider an analysis of the spatiotemporal utility of the events in a spatiotemporal region generated by tracking a spatial region across a unit of time. An analysis may divide it into two subregions R_1 and R_2 generated by tracking the elements of a twofold partition of the spatial region across the unit of time. Figure 5.5 depicts the subregions and uses a triangle to represent an event. The triangle event, which is partly in each subregion, may be conveniently assigned to

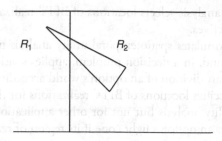

Figure 5.5 An event partially occurring in two regions

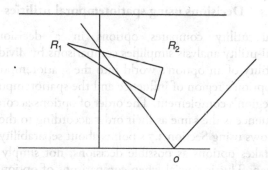

Figure 5.6 An option's light cone

the left subregion R_1 because it contains the last point of the event's realization. Using points of realization to locate events for a spatiotemporal-utility analysis involving a partition of a region into subregions prevents double counting events that occur partly in multiple subregions.

However, consider an analysis of the spatiotemporal utility of an option o's region of influence, a light cone that starts in the subregion R_2 and that Figure 5.6 depicts. The same triangle event, which overlaps the option's light cone, is now conveniently assigned to the option's region of influence, and so to subregion R_2 rather than to subregion R_1, because the option may influence the event by influencing part of the event. Perhaps the event would not occur if the option were not realized because the part in the option's region of influence would not occur if the option were not realized.

The best assignment of events to regions depends on the spatiotemporal-utility analysis's objectives. An analysis of the spatiotemporal utility of events in a region works provided that its assignment of events to subregions puts each BIT the region realizes into exactly one subregion. An apt assignment for an analysis selects locations of BITs' realizations to advance the analysis's objectives.

This chapter formulates spatiotemporal-utility analysis using locations of BITs' realizations and, in a decision problem, applies spatiotemporal-utility analysis to a two-part division of an option's world according to Section 5.1's conventions. It specifies locations of BITs' realizations for this application of spatiotemporal-utility analysis but not for other applications and so puts a BIT's realization in an option's light cone if its region of realization overlaps the light cone.

5.5   Decisions using spatiotemporal utilities

Spatiotemporal utility compares options in a decision problem. Spatiotemporal-utility analysis simplifies comparisons by dividing the spatiotemporal utility of an option's world into the spatiotemporal utility of events in the option's region of influence and the spatiotemporal utility of events in the region's complement. The order of options according to their regions of influence is the same as their order according to their worlds, as this section shows using Section 3.7's points about separability.

Chapter 2 takes options as possible decisions, not simply as acts that might be chosen. This is crucial when comparisons of options use spatiotemporal utility. A decision to perform an act may have consequences that the act itself lacks. The difference in consequences does not matter for a comparison of options according to their worlds because a choice's world and the act chosen's world are the same, assuming that the choice and act are indivisible. However, the difference matters for a comparison of options according to their results, that is, the events in their regions of influence. The choice and act chosen have different regions of influence. Because of the difference, an evaluation of a choice's region of influence may disagree with an evaluation of the chosen act's region of influence. For example, a child's decision to jump off the high diving board may cause a period of anxiety that the jump itself does not cause. Then, the decision to jump may have a lower spatiotemporal utility than does the jump itself.

This section assumes that all the options in a decision problem have the same time of realization. That is, some moment is each option's moment of realization if realized. Similarly, the options have the same place of realization, the decider's location, which the section's normative model takes as a point. The section assumes that each option is a point-event and if realized occupies the same space-time point as any other option. Granting that options are the decisions an agent may make at a particular time and place, the assumption of a common time and place of realization for options is a small idealization. Being possible decisions, the options that resolve a decision problem have roughly the same tiny spatiotemporal location, and the idealization just makes their location the same space-time point. This assumption allows comparison of options to compare the contents of the options' regions of influence, each of which begins at the same space-time point, so that the options' regions of influence coincide (given that no option affects the speed of causation). The assumption ensures that comparisons an analysis reaches by evaluating events in options' regions of influence are the same as those it reaches by evaluating their worlds. The events in those regions settle the options' comparisons and an option's choiceworthiness.

A spatiotemporal-utility analysis of an option's world separates the world's evaluation into an evaluation of its components. Then, comparison of options may eliminate components that are the same for all options. A ranking of options using the spatiotemporal utility of their results is the same as their ranking according to comprehensive utility. A demonstration of the rankings' equivalence uses Section 3.7's conclusion that any type of utility obtained by removing consideration of the same basic intrinsic attitudes from evaluations of options according to comprehensive utility yields the same comparisons of options as does comprehensive utility. The demonstration first separates evaluation of an option's world into evaluation of events in the option's region of influence and evaluation of other events common to all options' worlds and, second, puts aside evaluation of the other events. The switch from comprehensive to spatiotemporal utility for comparison of options is equivalent to adopting a new zero point for the options' utility scale. Changing a utility scale's zero point does not affect its ranking of options, so spatiotemporal utility ranks options just as comprehensive utility does.

The demonstration of agreement of rankings shows that comparisons of options using spatiotemporal utility agree with comparisons of options using comprehensive utility given removal of a common part of the options' worlds – a common block of events. It uses spatiotemporal-utility analysis to divide the spatiotemporal utility of an option's world, that is, the spatiotemporal utility of the world's region of realization – the world's spatiotemporal span. The analysis partitions the world's region of realization into the option's region of influence and its complement in the option's world.

For an option o, the option's world is $W[o]$. Given complete information, $SU(W[o]) = SU(R[o]) + SU(-R[o])$, where $-R[o]$ expresses the events in the complement of o's region of influence in $W[o]$. To accommodate incomplete information about an option's world, the formula's generalization introduces (1) $O[o]$, the proposition that o's world obtains; (2) $R'[o]$, the proposition that o's result obtains; and (3) $-R'[o]$, the proposition that the events in the complement of o's region of influence obtain. The quantities in the general formula estimate their counterparts in the formula for the case of complete information. $SU(O[o])$ is an estimate of $SU(W[o])$, $SU(R'[o])$ is an estimate of $SU(R[o])$, and $SU(-R'[o])$ is an estimate of $SU(-R[o])$. The general formula states that $SU(O[o]) = SU(R'[o]) + SU(-R'[o])$.

If an event is outside any option's region of influence, it is outside every option's region of influence. That is, it belongs to the options' common background. Therefore $SU(-R[o])$ and its estimate $SU(-R'[o])$ are constant for every option in a decision problem. That the events outside an option's region of influence are the same for every option follows from two assumptions of spatiotemporal-utility analysis, given that events are identified by sets of point-events. According to the first assumption, the regions of influence of all options in a decision problem coincide; the regions' content varies from option to option, but the regions are the same and so have the same complements. This assumption holds because options are possible decisions with the same time and place of realization, and the analysis makes their regions of influence as broad as necessary to ensure inclusion of every point where any option may causally influence events in any world of positive probability that realizes the option. The analysis identifies events outside the influence of a decision problem's resolution by reviewing the decision problem's entire set of options. According to the second assumption, the events in an option's region of influence are the events whose regions of realization intersect the option's region of influence. This follows from the analysis's conventions for locating events.

By the first assumption, every event that any option's realization might affect falls within the options' common region of influence. Events outside their region of influence, and in its complement, are the same from one option's world to another option's world. The constancy of events outside options' regions of influence holds for point-events and extended events alike because, by the second assumption, the assignment of every event to either an option's region of influence or its complement depends only on whether the event's region of realization intersects the option's region of influence. No event belonging to any option's region of influence belongs to the common complement of options' regions of influence. The two assumptions entail that the events in the complement of an option's region of influence are the same for all options.

An option's comprehensive utility evaluates the option's outcome. It is an estimate of the comprehensive utility of the option's world. The world's region of realization is the world's spatiotemporal span. So, $U(o) = U(O[o])$ $= SU(O[o])$. Hence, $U(o) = SU(O[o])$. Given that $SU(O[o]) = SU(R'[o]) +$ $SU(\ R'[o])$, it follows that $U(o) = SU(R'[o]) + SU(-R'[o])$. Because $SU(-R'[o])$ is constant for all options, a switch from $U(o)$ to $SU(R'[o])$ for evaluation of options amounts to subtraction of a constant $SU(-R'[o])$ from $U(o)$. In other words, $SU(R'[o]) = U(o) - SU(-R'[o])$, where the subtrahend evaluates the status quo. Moving from $U(o)$ to $SU(R'[o])$ for all options subtracts the same spatiotemporal utility $SU(-R'[o])$ from each option's comprehensive utility. Because evaluations of options using U and SU differ only by the zero point for their utility scales, the options' ranking according to SU is the same as according to U.

Given the chapter's conventions for assignment of events to an option's region of influence, each event in an option's world belongs either inside or outside the option's region of influence, and not both inside and outside this region. Hence, a spatiotemporal-utility analysis of the option's world using the option's region of influence does not omit or double count events in the option's world. If a BIT's realization in the option's world straddles the boundary of an option's region of influence, spatiotemporal utility's conventions assign it to the option's region of influence.

In some cases, a BIT's realization is assigned to an option's region of influence or to its complement, although it is not entailed by the point-events in the option's region of influence or by those in its complement. A BIT whose region of realization straddles the boundary of the option's region of influence has this feature. The point-events of the region to which it is assigned entail only its partial realization. Suppose that the point-events in an option's world $W[o]$, but not either the point-events in the option's region of influence or the point-events in its complement, entail some BIT's realization. However the BIT's realization is assigned, in a case of complete information, it may turn out that either $SU(R[o]) \neq IU(R[o])$ or $SU(-R[o]) \neq IU(-R[o])$ because intrinsic utility goes by entailment. Nonetheless, $SU(W[o]) = SU(R[o]) + SU(-R[o])$ because, according to spatiotemporal utility's conventions, a specification of events in an option's region of influence and a specification of events outside the option's region of influence together specify all the events in the option's world.

Spatiotemporal-utility analysis using a proposition's region of influence and its complement also applies to a conditional spatiotemporal utility. The spatiotemporal utility of an option's world given a state equals the spatiotemporal utility of events in the option's region of influence given the state plus the spatiotemporal utility of events in the region's complement given the state. That is, $SU(W[o]$ given $s) = SU(R[o]$ given $s)$ + $SU(-R[o]$ given $s)$. Because the world's region of realization is the world's span, the left side of the equation has the same value as the comprehensive utility of the option's world given the state. That is, $U(W[o]$ given $s) = SU(W[o]$ given $s)$. The second summand assesses events that obtain given the state independently of the option adopted. It is the same for all options because the complement of each option's region of influence is the same and has the same content. So, its subtraction to obtain the first summand alone amounts to a change in the zero point of the scale for conditional comprehensive utility. Given incomplete information, generalized equations govern the spatiotemporal utilities that estimate the quantities in the equations for the case of complete information. That is, $SU(O[o]$ given $s) = SU(R'[o]$ given $s) + SU(-R'[o]$ given $s)$, and $U(O[o]$ given $s) = SU(O[o]$ given $s)$. $SU(R'[o]$ given $s)$ estimates the spatiotemporal utility of events in o's region of influence given s, and similarly for $SU(-R'[o]$ given $s)$.[2]

Consequently, expected-utility analyses rank options the same way using either conditional spatiotemporal utilities or conditional comprehensive utilities. Section 2.5's formula for expected-utility analysis applied to an option o's comprehensive utility states that $U(o) = \sum_i P(s_i$ given $o)$ $U(o$ given $s_i)$. The formula applied to the spatiotemporal utility of the option's result states that $SU(R'[o]) = \sum_i P(s_i$ given $o)SU(R'[o]$ given $s_i)$, granting that $P(s_i$ given $R'[o]) = P(s_i$ given $o)$ because o occurs if and only if $R'[o]$ does. Because $U(o$ given $s_i) = U(O[o]$ given $s_i)$, the previous paragraph's equations establish that $U(o$ given $s_i) = SU(O[o]$ given $s_i) = SU(R'[o]$ given $s_i) + SU(-R'[o]$ given $s_i)$. The second summand estimates the spatiotemporal utility of events in the complement of o's region given s_i. This quantity is the same for every option and state. So, subtracting it to obtain $SU(R'[o]$ given $s_i)$ alone is equivalent to changing the zero point of the scale for the conditional comprehensive utility of an option-state pair. Hence, the expected-utility ranking of options obtained using $U(o$ given $s_i)$ is the same as the ranking obtained using $SU(R'[o]$ given $s_i)$.

---

[2] $SU(R'[o]$ given $s) = \sum_j P(r_j$ given $s)IU(r_j$ given $s)$, where r_j ranges over possible full specifications of the events that *would be* in o's region of influence if o *were* realized given that s *is* the case. Section 2.5 reviews the distinction between indicative and subjunctive supposition and explains the type of supposition a conditional utility uses for its condition.

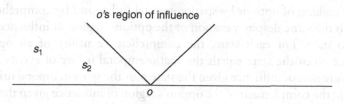

Figure 5.7 Independent states

5.6 Independence

Suppose that in expected-utility analyses of the options in a decision problem, the states are independent of the options in the sense that the states are outside the options' common region of influence. Figure 5.7 illustrates this type of independence in a case in which the options' common light cone and so option o's light cone form their common region of influence, and the states outside it are s_1 and s_2. Independence in this sense entails that the states are causally independent of the options and that in applications of causal decision theory a state's causal probability given an option equals the state's probability. Given this independence, the options' rankings according to comprehensive and spatiotemporal utility agree even after weakening assumptions about the options' region of influence.

Section 5.5's method of ranking options according to spatiotemporal utility assumes that options have the same region of influence in each state. Given that states are independent of options, the method may allow a state to affect the options' region of influence without upsetting the ranking of options. Instead of specifying the same region of influence for all options given all states, a ranking of options according to spatiotemporal utility may allow the options' region of influence taken with respect to one state to differ from their region of influence taken with respect to another state.

Imagine that in a decision problem information about the options' region of influence suggests the region's dependence on states. Supposing a state may resolve uncertainty about the options' region of influence. Given the supposition, their region may be narrower than a region broad enough to accommodate all uncertainty about the extent of the options' influence. Given independence, instead of assuming that for all options in all states, the options' regions of influence coincide, it suffices to assume that given each state, the options' regions of influence coincide. The common region's contraction given a state reduces the scope of spatiotemporal utility's evaluations of options given the state. This simplifies deliberations.

The ranking of options by spatiotemporal utility and by comprehensive utility is the same despite variation of the options' region of influence from state to state. For each state, the comprehensive utility of an option's outcome given the state equals the spatiotemporal utility of events in the option's region of influence given the state plus the spatiotemporal utility of events in the complement of the option's region of influence given the state. The second summand is the same for all options because the state is outside every option's region of influence. So, for each state, a constant exists such that the comprehensive utility of any option given the state is the same as the spatiotemporal utility of events in the option's region of influence given the state, after the constant's addition to the spatiotemporal utility. Because, given independence, a state's probability is the same with respect to each option, adding a constant to the spatiotemporal utility of events in each option's region of influence given the state adds a constant to the non-conditional spatiotemporal utility of events in each option's region of influence. The addition is equivalent to changing the zero point of the scale for an option's conditional spatiotemporal utility given the state and so also for an option's nonconditional spatiotemporal utility, an expected spatiotemporal utility. Given the assumption that states are independent of options, adding the same number to the spatiotemporal utilities of the events in the options' region of influence given a state simply adds the same probability-weighted spatiotemporal utility to the expected spatiotemporal utilities of the events in the options' region of influence. It does not change the options' expected spatiotemporal-utility comparisons. Similarly, the options' expected spatiotemporal-utility comparisons do not change after adding a constant to the spatiotemporal utilities of events in the options' region of influence given any other state. So, the shift from comprehensive utility to spatiotemporal utility produces no change in options' utility comparisons.[3]

Spatiotemporal-utility analysis, formulated using spatiotemporal utility, explicates a common type of spatiotemporal separation that justifies simplification of choices. The separation allows deliberations to fix a spatiotemporal region and evaluate events in other spatiotemporal regions. In particular, deliberations may evaluate events in the options' region of

---

[3] Jeffrey ([1965] 1990: chap. 2) examines minimizing expected regret. Regret has a reference point that shifts from state to state. Nonetheless, minimizing expected regret is equivalent to maximizing expected utility if states are independent of options. Similarly, given that a state affects the options' region of influence, spatiotemporal utility has a reference point that shifts from state to state. Nonetheless, maximizing spatiotemporal utility is equivalent to maximizing comprehensive utility if states are independent of options.

influence while fixing and ignoring its complement. Therefore, returning to the chapter's opening example, a homeowner deciding about a remodeling project justifiably evaluates the project by focusing on events the remodeling may affect.

This chapter shows how to evaluate an option in a decision problem using only the basic intrinsic attitudes realized in an option's region of influence according to conventions for locating their realizations. Chapter 6 uses causal relations to trim further the considerations an option's evaluation processes. Nonetheless, in special cases, Chapter 5's methods are easier to apply than are Chapter 6's methods because identification of relevant events in an option's region of influence is easier than identification of an option's relevant consequences. Also, Chapter 5's methods make fewer metaphysical assumptions than do Chapter 6's methods. This parsimony appeals to decision theorists, such as Richard Jeffrey ([1965] 1990: sec. 10.4), who favor reducing decision theory's reliance on causal relations.

5.7 Appendix: Consistency

A world is spread through space and time. A set of possible worlds adds another dimension to space and time. A possible event occurs at a place, time, and world. A gamble produces a chance for an event located at a place and a time in a possible world. Possible outcomes of an option's realization have as parts realizations of BITs at places, times, and worlds. Different ways of combining chances of their realizations yield different methods of calculating the option's comprehensive utility. This appendix verifies that calculations of an option's comprehensive utility that use spatiotemporal utilities are consistent with canonical calculations.

Section 5.5 observes that for an option o, $U(o) = U(O[o]) = SU(O[o])$. A demonstration shows that a canonical analysis of $U(o)$ agrees with a spatiotemporal analysis of $SU(O[o])$ so that comprehensive and spatiotemporal utility evaluate options consistently. The demonstration transforms the spatiotemporal-utility analysis of $SU(O[o])$ into an equivalent, canonical spatiotemporal-utility analysis. Then, it shows that the canonical spatiotemporal-utility analysis of $SU(O[o])$ agrees with the canonical analysis of $U(o)$.

According to Section 5.3's presentation of spatiotemporal-utility analysis,

$$SU(p) = \sum_i P(s_i \text{ given } p) \sum_j SU(p_j \text{ given } s_i)$$

where s_i ranges over the elements of a partition of states, and p_j ranges over specifications of the events in the elements of a partition of p's region of realization given s_j. Using worlds as states yields a basic version of this analysis:

$$SU(p) = \sum_i P(w_i \text{ given } p)\sum_j SU(p_j \text{ given } w_i)$$

where w_i ranges over possible worlds that realize p. The general form of analysis of $SU(p)$ using states follows from this form using worlds, according to the methods of proof in Weirich (2001: sec. A.2).

A canonical spatiotemporal-utility analysis of $SU(p)$ transforms the foregoing analysis using worlds by applying Section 5.2's definition of conditional spatiotemporal utility. According to the definition, $SU(p \text{ given } q) = \sum_i P_q(r_i \text{ given } p)\sum_j IU(\text{BIT}_{ij})$, where r_i ranges over possible full specifications of p's region of realization given q, and BIT_{ij} ranges over the objects of BITs realized in p's region of realization according to r_i. Suppose that p is a full specification of events in its region of realization. Then, given p, r_i has positive probability given q if and only if $r_i = p$. Moreover, given the equality, $P_q(r_i \text{ given } p) = 1$. In this case, $SU(p \text{ given } q) = \sum_i IU(\text{BIT}_i)$, where BIT_i ranges over the objects of BITs realized in p's region of realization given q. Therefore, because p_j is a possible full specification of events in its region of realization, $SU(p_j \text{ given } w_i) = \sum_k IU(\text{BIT}_{jk})$, where BIT_{jk} ranges over the objects of BITs realized in p_j's region of realization given w_i. These are the BITs whose realizations p_j entails. A world w_i where p is realized includes a specification of events in p's region of realization, so all the probabilities arising in the summation for $SU(p_j \text{ given } w_i)$, having w_i as a condition, are 1 or 0 according as p_j is compatible or incompatible with w_i. As a result, $SU(p_j \text{ given } w_i) = \sum_k IU(\text{BIT}_{ijk})$, where BIT_{ijk} ranges over objects of BITs assigned to the jth element of the partition of p's region of realization given w_i. Using this identity for the conditional spatiotemporal utility reduces the spatiotemporal-utility analysis of $SU(p)$.

$$SU(p) = \sum_i P(w_i \text{ given } p)\sum_j \sum_k IU(\text{BIT}_{ijk})$$

where BIT_{ijk} ranges over the objects of the BITs in the jth element of the partition of p's region of realization given w_i. This is the canonical version of spatiotemporal-utility analysis.

Applying a canonical spatiotemporal-utility analysis to $SU(O[o])$ yields the equality

$$SU(O[o]) = \sum_i P(w_i \text{ given } o)\sum_j \sum_k IU(\text{BIT}_{ijk})$$

where BIT_{ijk} ranges over objects of the BITs in the jth element of a partition of w_i's region of realization given w_i. This follows because (1) $P(w_i \text{ given } O[o]) = P(w_i \text{ given } o)$, (2) $P(w_i \text{ given } o)$ has positive probability only when w_i

is compatible with *o*, and (3) $O[o]$'s region of realization is w_i's region of realization when w_i is compatible with *o*.

Each BIT realized in w_i is assigned to just one element of a partition of w_i's region of realization. Therefore the canonical analysis of $SU(O[o])$ implies that $SU(O[o]) = \sum_i P(w_i \text{ given } o) \sum_{j \in \{ij\}} IU(BIT_j)$, where $\{ij\}$ indexes the BITs realized in w_i. A canonical analysis of $U(o)$ using expected-utility analysis and intrinsic-utility analysis also says that $U(o)$ equals this sum. Therefore, $U(o) - SU(O[o])$, as consistency requires.

CHAPTER 6

Causal utility

People making choices consider the consequences of their choices. A shopper buying a new car considers the consequences of her purchase for her transport and her budget. Evaluations of options that focus on consequences are efficient because they put aside irrelevant considerations. This chapter makes precise and justifies evaluation of options by their consequences.

Evaluating options according to events in their regions of influence, as discussed in Chapter 5, ignores only, but not all, events common to all options' realizations. For efficiency, evaluation of options should ignore all such events, including events occurring in every option's region of influence. Causal utility achieves this simplification.[1]

Because, as the book's Introduction notes, other works cogently argue for causal decision theory, this chapter assumes the theory and uses basic intrinsic attitudes to streamline the theory's evaluations of options. The chapter defines an option's consequences, explains how causal reasoning assigns a probability to an option's having a specific consequence, introduces causal utility and causal-utility analysis, and justifies using an option's causal utility to simplify the option's evaluation. It justifies making choices simpler by evaluating only options' consequences.

6.1 Consequences

Introductions of utility often present it as an evaluation of the possible consequences of a proposition's realization. They take the utility of having fire insurance on a house, for example, to be an evaluation of the consequences

[1] For helpful comments on the material in this chapter, I thank audiences at the University of Pittsburgh, the 2012 Editors Conference of the *British Journal for the Philosophy of Science* (especially my commentator, Steven French), the University of Sydney, the Australian National University, and the 2012 Central States Philosophical Association Meeting.

of having fire insurance if there is a fire, and the consequences of having fire insurance if there is no fire. For simplicity, textbooks often take consequences in such examples to be monetary. Given this simplification, the consequences of having fire insurance if there is a fire are the insurance's cost and the compensation the insurance provides for fire damage. The consequences of having fire insurance if there is no fire are the insurance's cost.

A type of utility, comprehensive utility, evaluates the possible outcomes of a proposition's realization. A proposition's *outcome*, in the comprehensive sense I adopt, is everything that would be true if the proposition were true. It is a possible world, which a maximal consistent proposition represents. A proposition's outcome is an element of the set of worlds that make the proposition true. The outcome of having fire insurance if there is a fire includes the fire itself as well as the net compensation the insurance provides for fire damage. It includes the consequences of having the insurance if there is a fire, but also everything else that would be true if one had fire insurance and there is a fire. It includes even events prior to the insurance's purchase that having fire insurance does not influence.

Evaluations of decisions to buy insurance and to not buy insurance typically do not examine the options' total outcomes. They use a type of utility, causal utility, with scope narrower than comprehensive utility's scope, to focus on the options' consequences. This chapter explicates and justifies common decision methods that, for efficiency, use causal utility. As in other chapters, it treats ideal agents with quantitative probability and utility assignments.

An event's consequences are notoriously difficult to specify precisely. Is Caesar's death a consequence of Brutus's stabbing him even though many others also stabbed him? Is World War II a consequence of Adolph Hitler's birth even though many other events were necessary for the war's outbreak? Does stealing the water of someone hiking in the desert have his death as a consequence even if his water was poisoned? Does my neighbor's not watering my plants, when I fail to water them, have my plants' death as a consequence? An account of consequences faces many hard questions. Earlier chapters put aside these questions to simplify applications of utility analysis. Comprehensive utility evaluates outcomes rather than consequences to skip separating the consequences of an option's realization from the other events that would obtain if the option were realized. Assessing outcomes (and, similarly, assessing futures and regions of influence) bypasses identification of consequences.

In one sense, it is simpler to evaluate all that would happen if an option were realized than to evaluate the option's consequences. It is conceptually simpler because the evaluation does not sort events using the consequence relation. However, it is more complicated in another sense. An evaluation has to consider more events if it considers all that would happen if an option were realized than if it considers just the option's consequences. The evaluation is operationally more complicated. Some events in the outcomes of the options in a decision problem are irrelevant to deliberations. Events prior to adoption of an option are common to all the options' outcomes. So are events contemporaneous with or subsequent to adoption of an option but independent of the option adopted. Because the shared features of options' outcomes do not influence the options' ranking according to comprehensive utility, it is operationally efficient to put them aside and evaluate options using only their possible consequences.

Causal utility, which Section 6.7 defines, evaluates options putting aside the common part of their outcomes. It focuses on their consequences. Its introduction adopts an understanding of consequences that serves an account of evaluation of options. To make the difference among options count as part of the difference among their consequences, it takes an option as a consequence of itself. Then options' evaluations differ only because of differences in the options' consequences. Given Section 6.4's distinction between an option's consequences and its effects, the stipulation does not imply that an option is its own effect; not all consequences are effects.

6.2   Causal decision theory

How does an option's evaluation attend only to the option's consequences, and what justifies the evaluation's narrow focus? Causal decision theory sketches some answers. It evaluates an option by considering the option's consequences, in contrast with evidential decision theory, which evaluates an option by considering the events for which the option provides evidence. Causal decision theory attends to causal relations, whereas evidential decision attends to evidential relations. Causation rather than correlation guides causal decision theory's advice.

Chapter 2 adopts causal decision theory in the course of presenting expected-utility analysis. The theory's basic formula for an option o's expected utility is $EU(o) = \sum_i P(w_i \text{ given } o) U(w_i)$, where w_i ranges over worlds that might be o's world. $P(w_i \text{ given } o)$ is not a ratio of nonconditional probabilities but a causal conditional-probability, as Section 2.1 explains. Causal decision theory recommends two-boxing in Newcomb's problem because of the option's good consequences, although one-boxing offers evidence of riches.

Gibbard and Harper ([1978] 1981) present causal decision theory and also some ideas about separation of an option's utility into parts to make evaluation of options efficient. They ([1978] 1981: 156) state that a possible outcome is a proposition that fully specifies what may, for all the agent knows, be the relevant consequences of o, and they compute an option's utility with respect to a set of possible outcomes that are exclusive and exhaustive, omitting possible outcomes to which the agent assigns probability zero. A full specification of an option's relevant consequences is incompatible with any other full specification, and such a specification receives probability zero if it is incompatible with the agent's knowledge. Although Gibbard and Harper do not justify calculating an option's utility, for comparison of options, using the option's possible outcomes defined in terms of its possible consequences instead of worlds that, epistemically speaking, may be the option's world, this chapter supports a version of their formula for an option's utility.

Gibbard and Harper ([1978] 1981: 167–68) make two suggestions for simplifying causal decision theory's comparison of options. The first suggestion, concerning evaluation of an option, takes a state s as an option o's consequence if and only if, if o were realized then s would obtain, that is, $o \,\square\!\!\rightarrow s$; and for some other option o', it is not the case that if o' were realized s would obtain, that is, $\sim(o' \,\square\!\!\rightarrow s)$. A state s in this definition may be an element of the set of consequences constituting a possible outcome in the definition of expected utility, or it may be the whole outcome. If the elements satisfy the definition, then the whole outcome does. A state s is *unavoidable* if and only if for every option, if the option were realized then s would obtain.[2] Hence, a state s is a consequence of o if and only if $(o \,\square\!\!\rightarrow s)$, and s is avoidable. $P(o \,\square\!\!\rightarrow s) = P(s$ is a consequence of $o) + P(s$ is unavoidable) because the right-hand side of the equality has probabilities of incompatible propositions that exhaust the ways $(o \,\square\!\!\rightarrow s)$ may be true. Hence, $U(o) = \sum_s P(o \,\square\!\!\rightarrow s)U(s) = \sum_s P(s$ is a consequence of $o)U(s) + \sum_s P(s$ is unavoidable$)U(s)$, where s ranges over the elements of a partition of o's possible outcomes.

This separation of $U(o)$ using probabilities is not effective, however. $P(s$ is a consequence of $o) = P(o \,\square\!\!\rightarrow s)$ given that s is a possible outcome in Gibbard and Harper's sense and so specifies a possible set of o's relevant consequences. Hence, the second factor in the sum for $U(o)$ equals zero, so

---

[2] Horty (2001: sec. 4.4) defines the independence of a proposition's truth from an agent's acts. His definition resembles Gibbard and Harper's definition of an unavoidable outcome.

$\sum_s P(o \,\square\!\!\rightarrow s)U(s) = \sum_s P(s$ is a consequence of $o)U(s)$. Therefore, no separation occurs.

Gibbard and Harper's second suggestion about efficiency makes a comparison of two options o and o' evaluate each option using only consequences that they do not share. Each option's evaluation may put aside states that are consequences of both options. Consider an evaluation of the option o in its comparison with the other option o'. By probability's law of additivity, $P(o \,\square\!\!\rightarrow s) = P([o \,\square\!\!\rightarrow s] \,\&\, {\sim}[o' \,\square\!\!\rightarrow s]) + P([o \,\square\!\!\rightarrow s] \,\&\, [o' \,\square\!\!\rightarrow s])$. Hence, $U(o) = \sum_s P(o \,\square\!\!\rightarrow s)U(s) = \sum_s P([o \,\square\!\!\rightarrow s] \,\&\, {\sim}[o' \,\square\!\!\rightarrow s])U(s) + \sum_s P([o \,\square\!\!\rightarrow s] \,\&\, [o' \,\square\!\!\rightarrow s])U(s)$, where s ranges over the elements of a partition of o's possible outcomes. This formula separates o's utility into a part involving possible outcomes it shares with o' and a part involving possible outcomes it does not share with o'.

The separation is not effective, however. In a typical case, the possible outcomes of o all differ from the possible outcomes of a different option o'. A possible outcome of o and a possible outcome of o' differ in at least one consequence, so $P([o \,\square\!\!\rightarrow s] \,\&\, [o' \,\square\!\!\rightarrow s]) = 0$ for all s. In this case, $\sum_s P(o \,\square\!\!\rightarrow s)U(s) = \sum_s P([o \,\square\!\!\rightarrow s] \,\&\, {\sim}[o' \,\square\!\!\rightarrow s])U(s)$. The second factor in the sum for $U(o)$ equals zero. No separation occurs.[3]

Besides failing to separate an evaluation of an option's consequences, and its distinctive consequences, from an evaluation of other events accompanying it, Gibbard and Harper's suggestions about efficiency do not specify the scope of the utility function U. If it is comprehensive, then when applied to a possible set of consequences, it evaluates the world in which those consequences occur. $U(s)$, if it is comprehensive utility, evaluates s's world. Hence, it does not put aside irrelevant considerations to promote efficiency. Without scope restrictions, $U(s)$, their formula's atom of utility, equals the comprehensive utility of a possible world. The utility of an option's consequences and the utility of an option's nonconsequences each equal the utility of the option's world. The two utilities do not sum to the utility of the option's world.

This chapter reformulates Gibbard and Harper's proposals about efficiency so that they overcome the problems noted. An option's efficient evaluation separates an option's utility using the BITs that would be realized if the option were realized. It divides these BITs according to whether their realizations are consequences of the option or are unavoidable and uses

---

[3] The possible outcomes of two options differ in typical cases because outcomes are finely individuated with respect to all relevant matters, as Section 2.4 explains.

intrinsic-utility analysis to calculate the intrinsic utility of each group of BITs. This separation using intrinsic utilities makes effective the separation that Gibbard and Harper suggest.

Also, a revision using intrinsic utilities of realizations of BITs makes effective Gibbard and Harper's proposal to efficiently compare two options by putting aside consequences they share. It divides BITs that an option realizes into BITs that the option, but not its rival, realizes, and BITs that the option and its rival alike realize. Using this division of BITs makes pair-wise comparison of options more efficient than using the division of BITs that grounds an evaluation of each option according to consequences that the option does not share with all other options. It conducts each pair-wise comparison using the smallest set of BITs the comparison requires. Pair-wise comparison of all options generates a preference ranking of options and a choice. A general pair-wise procedure yields a ranking of all options more efficiently than does a general procedure evaluating each option with respect to all its consequences.

For example, imagine a graduate's choice among careers in law, medicine, and boxing, assuming BITs toward wisdom, satisfaction, and health. Suppose that the law offers wisdom, satisfaction, and health. Medicine offers wisdom and health. Boxing offers satisfaction. A comparison of the law and medicine may skip wisdom and compare satisfaction combined with health to health alone. The law wins. A comparison of the law and boxing may skip satisfaction and compare wisdom combined with health to their absence. The law wins and so wins overall. Elimination of shared consequences, defined using BITs, simplifies comparisons. Efficient comparison of all options evaluates options two at a time with respect to their distinctive consequences instead of evaluating all options with respect to all their consequences.

6.3   Causation

This section briefly reviews causation and also states, but does not argue for, the chapter's assumptions about causation, including empirical assumptions ruling out backward causation and causation faster than the speed of light. The section treats only points about causation that advance the chapter's objective of separating evaluation of an option's consequences from evaluation of other events. It skips over many issues concerning causation and, in particular, does not attempt a reductive account of causal relations. Causal-utility analysis is compatible with many accounts of causation.

Causation is a topic rather than a single relation. Many types of causal relations exist. Causal relations may hold between event-types or tokens, and between variables or values of variables. The relations may be physical, or they may be normative, involving responsibility for events. Being *the* cause of an event differs from being *a* cause of the event. Being the total cause of an event differs from being a partial, or contributing, cause of the event. The next section describes the causal relation that identifies an option's consequences.[4]

Causation and constitution differ, although in some cases they resemble each other. The components of a composite act do not cause the act's realization but rather constitute the act's realization. For example, the steps of a walk constitute the walk but do not cause it. Sometimes the distinction blurs. If you have $10 to spend on fruit and vegetables, does spending $5 on fruit cause you to spend no more than $5 on vegetables? Maybe it constitutes spending no more than $5 on vegetables. This section does not try to make the distinction precise.

An option's realization and its consequences' realizations are *events*. Events are the relata of the causal relation that identifies an option's consequences. Propositions represent events, and events are propositionally individuated, as I understand them. Poisoning the reservoir differs from pumping poisoned water into the reservoir, even if the poisoning and the pumping have the same physical realization. A death the poisoning causes differs from a painful death it causes, even if the two events have the same physical realization. Given incomplete information, the poisoning's two consequences may have different comprehensive utilities despite their physical equivalence.

Propositional individuation of events allows an event's nonoccurrence to count as an event. This accommodates cases in which not watering a plant kills it. Also, propositional individuation acknowledges events, such as a walk, that are spread across spatiotemporal regions. An extended event *e* is a partial cause of an extended event *e′* if some point-event belonging to *e* is a cause of some point-event belonging to *e′*. For example, suppose that an agent acts now to complete a sequence of acts begun in the past. The current act is a partial cause of the sequence's effects, although it is not a total cause of the sequence's effects.

---

[4] Dretske (1988: 42–45) distinguishes between enabling causes and triggering causes. Woodward (2003: 45–61) distinguishes direct, total, and contributing causes. Hall (2004) distinguishes causal production and causal dependence and holds that the latter guides rational decisions. Maslen (2004) holds that causation may be contrastive so that one event causes a second rather than a third. Hitchcock and Kobe (2009) study normative causal relations.

Propositional individuation of events makes events fine-grained in one sense. Events with the same physical realization may be distinct. Propositional individuation allows for events that are coarse-grained in another sense. Smith's pumping water is a more coarse-grained event than Smith's pumping water with his right hand. Although propositionally individuated, events may be coarse-grained because they allow multiple physical realizations. An event-type may have multiple realizations, whereas an event-token has just one realization. Yesterday's storm is an event-token with a duration. That a storm happens tomorrow is an event-type. Tomorrow's storm, if there is one, may have various durations. Variables may represent event-types, and their values may be event-tokens. However, the two values of an indicator variable for a BIT, the BIT's realization and its nonrealization, are not event-tokens of the same type.

An unrealized option is an event-type that a proposition individuates and that has multiple possible realizations. Options and their consequences may be unrealized and multiply realizable. There are many ways to place a bet, for example, and many ways of winning the bet. Between an option realized and its effect, a causal relation between event-tokens holds. Between an unrealized option and its unrealized effect, a different causal relation between event-types holds. The causal relation between event-types derives from causal relations between event-tokens in the option's world, the nearest world containing the option's realization.[5] In the option's world, a causal relation between the option and its consequences holds. To identify an option's consequences in a way that accommodates unrealized options, this chapter uses a causal relation between event-tokens in the option's world. An unrealized option has no consequences, but in the option's world it has consequences. An unrealized option is an event-type, but in the option's world the option's realization is an event-token.[6]

---

[5] For simplicity, this chapter assumes that a unique world exists such that it would be realized if the option were realized. In general, as Section 2.1's long note states, a counterfactual conditional is true just in case either its antecedent is impossible or some world in which its antecedent and its consequent are true is closer than any world in which its antecedent is true and its consequent is false.

[6] Actual events occur in the actual world. An event belongs to a world in which it happens. A possible event, taken as an event-type, is realized in a world and has a concrete event-token in the world. Lewis (1986b) recognizes unrealized possibilia. If an unrealized option were realized in just one possible world, or one favored world, it could be identified with its event-token in that world. An unrealized option is then a concrete possibilium, the realization it would have if it were realized. An unrealized event may be a possibilium realized in a nonactual possible world and having concrete features, such as a duration. Nonetheless, this chapter takes an unrealized option as an event-type to allow for an option's realization in multiple worlds, and even in multiple nearest worlds (although the chapter does not treat such cases).

To appraise an option's realization, causal decision theory attends to its possible consequences. The relevant possible consequences are objects of desire or aversion. They ground calculations of the option's expected utility. The *basic consequences* are realizations of BITs. Other consequences derive from realizations of BITs. Causal decision theory asks whether a BIT's realization is an option's consequence and not just whether it is correlated with the option (perhaps because of a common cause).[7] Which realizations of BITs are an act's consequences?

Causal responsibility is complex and in the law depends on identification of the salient cause of an event's production, a pragmatic issue. The type of causal responsibility that causal decision theory monitors is physical rather than pragmatic, and independent of norms. It directs pragmatic responses to decision problems and performs its role better if it is not itself affected by pragmatic considerations. Although causal decision theory treats agency, it considers an option's causal responsibility in a purely physical rather than a partly pragmatic sense for a BIT's realization. Reviewing accounts of some physical relations of causal responsibility sets the stage for the next section's interpretation of an option's consequences.

Lewis ([1973] 1986a) proposes an analysis of the relation of one event's being a cause of another distinct event. An event C has as a causal dependent an event E if and only if $\sim C \,\square\!\!\rightarrow \sim E$, where the counterfactual conditional is not backtracking. C is a cause of E if and only if the ancestral of causal dependence relates C and E; that is, C begins a sequence of causally dependent events that ends with E. This analysis targets causes and effects that have occurred, and an extension accommodates unrealized options and their possible consequences.[8]

---

[7] According to the theory, correlation with a BIT's realization may affect an option's utility if features of the BIT's realization are also the objects of other BITs whose realizations are the option's possible consequences. This case does not arise, however, because a BIT's realization does not have as features the objects of other BITs. Such features, were they to exist, would, by providing reasons for holding the intrinsic attitude, make it nonbasic.

[8] Sartorio (2013) holds that Lewis's account of causation fails to correctly characterize the way that causes are difference makers. It rules that one event is not a cause of another event if the second event would have occurred in the absence of the first. In place of Lewis's requirement of difference making, she advances another: "[I]t is part of the essence of causation that something cannot be a cause (thus, it cannot be part of the actual causal sequence) unless its absence would not have been a cause (thus, unless its absence would not have been part of the actual causal sequence)" (211). This weaker requirement of difference making allows Suzy's throwing a rock to cause a window's shattering even if in the absence of her throw Billy's throwing a rock would have shattered the window. Problems arise, however, because whether an event's absence would have been a cause of another event depends on how its absence would be realized. Suppose that a logic student takes the next course in the logic sequence because the instructor gave her an "A" for the course she just finished. If the instructor had not given her an "A", the instructor would have given her a "B", and she would have taken the next

The manipulation account of causation, although not an analysis because manipulation is itself causal, accommodates unrealized events. According to it, causes are toggles one manipulates to obtain effects. Adherents are Hausman (1998) and Woodward (2003). The manipulation account uses causal relations between variables having events as values to ground causal relations between events, as in Halpern and Pearl (2005).

Kutach (2012) proposes that in physics, the events in an event's future-directed light cone constitute the event's total effect, and the events in the event's past-directed light cone constitute its total cause. Despite his view's attractiveness, this chapter does not adopt it to identify an option's consequences because it makes an evaluation of an option's consequences as broad as an evaluation of the option's region of influence. Causal utility aims to minimize an option's consequences and so excludes some elements of an option's future-directed light cone. It excludes events common to the future-directed light cones of all the options in a decision problem because they do not differentiate the options' utilities.

Some causal relations are probabilistic and relate event-types: a flu shot lowers the physical probability of catching the flu. The probabilistic causation obtains, although an instance of the cause does not invariably produce an instance of the effect. In fact, in unusual cases with adverse susceptibility, a flu shot may provoke the flu. In general, if C causally influences E, then C raises the probability of E given control for background factors. This account of probabilistic causation does not qualify as a reductive analysis if the relevant background factors have a causal characterization.

A theory of causation advances laws of causation such as the anti-symmetry of causal relations. Is causation transitive? That is, if one event causes a second event and the second causes a third, does the first event cause the third? Although plausible, the principle of transitivity has some apparent exceptions.

One region-event may causally influence a second region-event that begins before it. The second region-event in its early stages may causally influence a third region-event that entirely precedes the first so that the first does not causally influence the third. For instance, a child's birth influences the mother's life. The mother's life influences the childhood of her

course in the sequence anyway. Then, according to the new requirement, the instructor's giving the student an "A" does not cause her to take the next course because its absence would also cause her to take the next course. The weakened requirement is still too strong to allow the instructor's giving the student an "A" to cause her to take the next course. To overcome this problem, the account may distinguish between the instructor's not giving an "A" and its nearest realization. Then, it may claim that the instructor's not giving an "A" does not cause the student to take the next course, and so giving an "A" makes a difference.

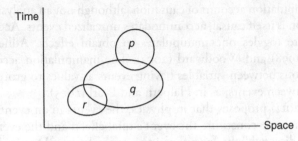

Figure 6.1 Causal influence's nontransitivity

schoolmates. But the child's birth does not influence the childhood of the mother's schoolmates. Figure 6.1 depicts region-events p, q, and r, such that p influences q, and q influences r, although p does not influence r.

For another example, suppose that a political candidate's speech affects the voting behavior of twins. It causes one twin to abstain but otherwise has no effect on voters. The twins' voting behavior affects the election's outcome because the voting twin's vote for the candidate is decisive. However, the candidate's speech did not affect the election's outcome. If the twin abstaining had voted, he would have voted for the candidate as his twin did.

Whether transitivity holds depends on the causal relation. Consider a causal relation defined in terms of light cones: p is a cause of q if and only if p is in q's past-directed light cone. According to this definition, if p is a cause of q, and q is a cause of r, then p is a cause of r. Transitivity holds for this causal relation. In contrast, counterfactual dependence, another relation in the causal family, is not transitive, as the following example shows: (1) If I had eaten my spinach, I would have been strong. (2) If I had been strong, I would not have eaten my spinach. (3) If I had eaten my spinach, I would not have eaten my spinach. It is possible for the first two conditionals to be true but not the last. Because the consequence relation, as the next section introduces it, rests on counterfactual dependence and admits region-events, it is not transitive.[9]

Next, this section addresses a worry that philosophers have about causation. Does causation exist? Hume debunks causation and takes its real counterpart as constant conjunction with contiguity and temporal

---

[9] Pearl (2000: 237) maintains that causation is transitive for events in chain-like causal structures, such as the falls of dominos that were standing in a row. Spohn (2001: 162) holds that causation is transitive. Hitchcock (2001) argues for the nontransitivity of some causal relations. Hall (2004) argues that Hitchcock's counterexamples to causation's transitivity touch only causal dependence and not causal production. Maslen (2004) defends the transitivity of causes.

precedence. Eating bread causes nourishment because with regularity nourishment follows eating bread.

An operationalistic decision theory eschews causation because it is not defined in terms of observables. Contemporary theories of meaning, however, allow implicit definitions of theoretical terms and do not insist on operational definitions. They allow a theoretical term's role in a theory's formulation to introduce the term. Nonoperational concepts make a theory less clear but may bring compensating gains in the theory's scope, explanatory power, and usefulness. Causation is understandable even if not clear in all cases. It is manifest in the realm of observables even if it does not have a precise definition in terms of observables. It has conceptual respectability without being operational. Moreover, evidence such as regularities supports its existence.

Also, reliance on comprehensive utility exclusively does not dispense with causality in utility analysis. It merely relegates causality to the outcomes of propositions evaluated. To calculate the comprehensive utility of reporting a criminal, say, a crime stopper considers whether, if he reports, he will be in a world where he has causal responsibility for apprehension of the criminal and so is entitled to a reward. Features of the option's outcome raise causal issues, even if the type of utility applied to the option is not causally defined. The gain from eschewing causal utility is not conceptual economy in the assessment of utility, but merely conceptual economy in the definition of utility. Furthermore, utility analysis relies on causally defined probabilities, even if it uses comprehensive instead of causal utilities. The resolution of Newcomb's problem, for instance, uses causal decision theory's causal conditional-probabilities. Hence, adopting comprehensive utility does not yield a net conceptual economy for utility analysis.

This chapter makes some assumptions about causation. First, single-case causation is fundamental. It is the ground of causal laws. Causation between event-tokens explains causation between event-types, although causation between event-types licenses inferences about causation between event-tokens. Second, free acts may have causes. Freedom and causation are compatible. Evaluation of decisions supposes changes in the agent's decision without any change in the agent's background beliefs and desires. Given the constant background, a change in the agent's decision may affect its evaluation for rationality. Evaluation keeps fixed background beliefs and desires when imagining realization of an option. It supposes that options are independent of them. However, the supposition is counterfactual; the background beliefs and desires may cause an option's realization. Third, an option's effects occur only in its future and in its light cone. This chapter's illustrations of causal-utility analysis take these empirical facts about causation as certain. The assumption grounds the illustrations but not the chapter's general account of causation. The general account allows for backward causation so that the past relative to a time may have events not in the past relative to a later time. The next section characterizes an option's consequences without empirical assumptions about causation's operation.

6.4   Consequences defined

Deliberations consider what would happen if an option were realized. Causal relations govern an account of what would happen. Causal utility evaluates an option by evaluating the option's *consequences*. It uses a causal relation to identify an option's consequences in a world with the option's realization. Which member of the family of causal relations identifies an option's consequences? This section describes the causal relation that best directs an option's evaluation according to its consequences. It provides causal utility's account of an option's consequences.

Causal decision theory does not take causation as correlation between event-types, for if it did, it would misdirect choice in decision problems such as Newcomb's problem. Because causal decision theory evaluates options in a decision problem, it uses a causal relation between events, taken as event-tokens rather than as event-types, to identify an option's consequences. For an unrealized option, it uses the option's causal relations to events-tokens in the option's world.

In an ordinary sense, an event has as a consequence any event that comes with it; an event's consequences are the event's outcome, result, upshot, sequel, or issue. An option's consequences in this sense include any event that would occur if the option were realized, or, more selectively, any event following the option's realization. This characterization is too broad for causal utility because it makes an option's consequences include every event, or at least every future event, that would occur if the option were realized. It makes causal utility's scope no narrower than temporal utility's scope.

An event's consequences in another common sense are the events that depend on the event. If this characterization takes dependence as counterfactual dependence, it is also too broad for causal utility because it makes an option's consequences include every event that would occur if the option were realized. Counterfactual dependence yields the nearest world in which the option occurs and so does not yield a selective account of an option's consequences.

A preliminary definition of an option's consequences states that an event c is a consequence of an option o if and only if $o \;\square\!\!\rightarrow c$ and $\sim\!o \;\square\!\!\rightarrow \sim\!c$. This definition eliminates events that happen regardless of the option's realization. However, a consequence is not just an event that occurs with the option and would not occur if the option were not to occur. Whether c would occur if o were not realized depends on which option would occur if o were not realized. If another option with the consequence c would be realized if o were not realized, the definition does not count c as o's consequence. If c is the case whether or not the agent were to adopt o, then the definition rules that c is not a consequence of o. It overlooks the possibility of a non-exhaustive pair of options that have a common consequence. For example, two ways of walking to school have arrival at school as a consequence. The definition does not count arrival as a consequence of either way of walking, but the evaluation of each option should count the event as a consequence. The preliminary definition gives an accurate account of an event's being an option's exclusive consequence but not an accurate account of an event's being an option's consequence *simpliciter*.

Gibbard and Harper ([1978] 1981: 167–68) take a state s to be a consequence of an option o if and only if $(o \;\square\!\!\rightarrow s)$ and $\sim\!(o' \;\square\!\!\rightarrow s)$ for some option o'. Couched in terms of events rather than states, the definition stipulates that an option's consequence is an event that would occur if the option were realized and is avoidable by realizing another option: an event c is a consequence of an option o if and only if $o \;\square\!\!\rightarrow c$ and $\sim\!(o' \;\square\!\!\rightarrow c)$ for some option o'. According to Gibbard and Harper's definition, an option's consequence is a counterfactually dependent event that does not follow from every option's realization. Their definition selects as an option's consequences the avoidable events in the option's world. It excludes events that occur regardless of the option realized but allows for two options having the same consequence.

Following Gibbard and Harper, this section says that an event is an option's *consequence* if and only if the event is avoidable and would occur if the option were realized. Because an option is avoidable and would occur if it were realized, the definition counts an option as a consequence of itself. This feature of the definition assists evaluation of an option by putting the option in the scope of its evaluation by its consequences.

The section's definition of an option's consequences, although serviceable, needs supplementation. It needs a supplementary account of the relation of comparative similarity that combines with the analysis of counterfactual conditionals to yield truth-values for the conditionals the definition uses. The next section furnishes a supplement that uses other causal relations to explicate the consequence relation. It makes the consequence relation non-backtracking given the empirical assumption that causation is forward.

Gibbard and Harper's definition of an option's consequences assumes that a unique world would be realized if the option were realized. When the assumption is not met, the conditionals the definition uses may lack truth-values (or have them only by stipulation). Taking the relation of being an option's consequence as a primitive concept that agrees with the definition when the conditionals have truth-values yields a generalization of the definition that dispenses with the assumption. This chapter's points need not generalize the definition of an option's consequences. For simplicity, they retain the assumption that for every option a unique nearest option-world exists. Under this assumption, if option o has the consequence c, c's avoidability given some alternative option o' is equivalent to c's not occurring if o' were realized.[10]

For an arbitrary contingent proposition, the consequences of the proposition's realization are avoidable events that would occur if the proposition were realized. Alternatives to the proposition's realization settle the avoidable events, and context settles the full set of alternatives. Realization of the proposition's negation is always an alternative to the proposition's realization. Events that would not occur if the proposition were not realized are avoidable (by processes that generate the proposition's realization rather than an alternative's realization). This chapter treats propositions that express options in a decision problem, and the alternatives to an option are the other options in the decision problem.

A propositional individuation of events permits an event that may have multiple realizations. In a decision problem, an unavoidable event may have a different realization given each option because each option

---

[10] Given that for every option a unique nearest option-world exists, subjunctive conditionals with an option o as antecedent obey the law of conditional excluded middle: for every event e, either $(o \ \square\!\!\rightarrow e)$ or $(o \ \square\!\!\rightarrow \sim e)$. Hence, $\sim(o' \ \square\!\!\rightarrow c)$ is equivalent to $(o' \ \square\!\!\rightarrow \sim c)$. When applied to cases that do not meet the uniqueness assumption, Gibbard and Harper's definition counts more events as consequences than it would if the second conjunct were $(o' \ \square\!\!\rightarrow \sim c)$. Its wide scope includes every relevant event. Whether it includes irrelevant events is controversial. Also, if an option increases an event's propensity to occur but does not ensure the event's occurrence, Gibbard and Harper's definition counts the change in the event's propensity, but not the event itself, as the option's consequence. A generalization of their definition may count the event too.

Some recent accounts of conditionals defend the law of conditional excluded middle. Bradley (2012) uses ordered sets of possible worlds to modify the usual possible-worlds semantics for conditionals. An ordered set for a conditional represents not merely the actual world but also the world the antecedent yields. The modifications accommodate indicative and subjunctive supposition of a conditional's antecedent, support the law of conditional excluded middle, and for indicative conditionals show that the probability of a conditional equals the corresponding conditional probability.

influences subtly events spread throughout its light cone, as a rock's fall into a pond influences subtly wavelets everywhere on the pond's surface. Suppose that a walker is deciding whether to shout a greeting to a distant friend. With respect to the decision problem, an imminent rain shower is an unavoidable future event, but the shower's realization differs slightly for the options because shouting creates airwaves that slightly affect the trajectories of raindrops. The rain shower is not any option's consequence, although each option's consequences include details of the shower's realization.

In a decision problem, are all an option's consequences in its region of influence? All avoidable events that counterfactually depend on the option are in its region of influence. If an event is avoidable, then for some option the event fails to occur given the option's realization. So, the event is not in the complement of the common region of influence of the options in the decision problem. Given its occurrence in the option's world, the event is at least partly in, and so in, the option's region of influence. Therefore, an evaluation that attends exclusively to consequences has a scope as least as narrow as spatiotemporal utility's scope.

As with some causal relations, the consequence relation is not transitive. In Section 6.3's case of the twins, one votes and the other abstains. Both twins' abstaining is an alternative act for the pair. Letting s stands for the candidate's speaking and v for the twins' voting behavior, $(s \: \square \!\!\rightarrow v)$ & $\sim(\sim s \: \square \!\!\rightarrow v)$, because neither twin would have abstained had the candidate not spoken. So, the candidate's speaking has their voting behavior as a consequence. Letting e stand for the candidate's election and v' for both twins abstaining, $(v \: \square \!\!\rightarrow e)$ & $\sim(v' \: \square \!\!\rightarrow e)$, because the voting twin was decisive. So the twins' voting behavior has the candidate's election as a consequence. However, because the candidate's not speaking would not have prevented his election, $(\sim s \: \square \!\!\rightarrow e)$. Therefore, it is false that $(s \: \square \!\!\rightarrow e)$ & $\sim(\sim s \: \square \!\!\rightarrow e)$. Assuming that the candidate's not speaking is the only alternative to his speaking, his speaking does not have his election as a consequence.

Are an option's being a cause of an event and the event's being a consequence of the option the same relation? No, an event's being an option's consequence is not the same as its being the option's effect. An option is its own consequence but is not its own effect, for instance. Also, suppose that a gambler rolls a die, and it comes up six. The gambler's rolling the die increases the physical probability of the die's coming up six but is not a cause of its coming up six. Its coming up six is an indeterministic event without any cause, granting that causes determine their effects.

Yet, it is avoidable. If the gambler had not rolled the die, it would not have come up six. Moreover, the die's coming up six counterfactually depends on the gambler's rolling the die (because both events occur), so it counts as a consequence of the gambler's rolling the die.[11]

Moreover, an event may be an effect of all options. In that case, it is not any option's consequence because it is unavoidable. First, consider the case of an assassin with a backup. If the assassin does not kill the victim, his backup will. Either the assassin shoots the victim or the backup assassin does. Whatever the assassin does, the victim dies. Because the victim's death is unavoidable, it is not a consequence of the assassin's shooting. Nonetheless, the victim's death is an effect of the assassin's shooting (and is not an effect of the assassin's not shooting). The assassin's deliberations about shooting may omit the victim's death despite its being an effect of the assassin's shooting. His evaluation of options may ignore it because it follows from all his options. It does not discriminate among his options even if it matters to him. Causal utility puts aside the victim's death, although it is an effect of the assassin's pulling the trigger.[12]

Second, suppose that late at night, a gangster in a doorway, whatever he does, either by moving or by staying still, will frighten a passing walker. This common effect of his options is unavoidable and so is not a consequence of any option.

Third, imagine that an agent has a BIT toward resolving a decision problem. Then, an option's realizing that BIT counts as a relevant effect of the option. Nonetheless, because all options have the same effect, it is unavoidable and so not a consequence of any option. Efficient evaluation skips common effects of options, so the definition of an option's consequences also puts them aside.

The foregoing cases show that an option's consequences are not quite the same as the option's effects. Not all of an option's consequences are its effects, and not all of an option's effects are its consequences. Defining an option's consequences as avoidable counterfactually dependent events distinguishes them from the option's effects. Causal utility reviews an option's consequences rather than its effects.

---

[11] Suppose that wearing a cap and gown at graduation constitutes rather than causes continuation of a tradition. Although continuation of the tradition is not the option's effect, it is the option's consequence because it is an avoidable event that counterfactually depends on the option.

[12] Although the victim's death is unavoidable, the assassin's killing the victim, a fine-grained consequence, is avoidable. If the assassin cares about responsibility for the victim's death, then his killing the victim is a relevant consequence of his shooting.

Can the consequence relation adjust to agree with causation? In the case of the assassin with a backup, suppose that the assassin shoots and the backup does not. To settle whether the victim's death is a consequence of the assassin's shooting, keep fixed the backup's not shooting. Then, if the assassin had not shot, the victim would not have died. Holding some events fixed, the victim's death counterfactually depends on the assassin's shooting and is avoidable.[13]

Decision theory prefers not adjusting the consequence relation so that an option's consequences agree with its effects. Unadjusted, the consequence relation suits an option's efficient evaluation. It identifies consequences in the sense that matters. An option's consequences in the relevant sense are the avoidable events that would occur if the option were realized.

Although an option's consequences may differ from the events of which the option is a cause, the consequence relation is in the causal family, and causal relations govern the comparative similarity relations that settle the truth-conditions of the subjunctive conditionals defining avoidability and counterfactual dependence, and so the consequence relation. Causal utility thus is an apt name for an evaluation of an option that attends exclusively to the option's consequences.[14]

6.5   Intervention

Subjunctive conditionals appear in the definition of an option's consequences, and causal decision theory's calculations of an option's expected utility attach probabilities to them. For an option o and state s, the truth conditions for the subjunctive conditional $o \,\square\!\rightarrow s$ presume similarity relations among possible worlds, for they say that the conditional is true if in the nearest option-world the state is true. This section explains how causal models settle these similarity relations by identifying causal relations that similar worlds share and how taking an option's realization as an intervention with respect to a causal model yields the nearest option-world.[15]

---

[13] An option is a consequence of itself but does not cause itself. To make the consequence relation agree with causation, one must restrict it to pairs of events such that the first does not entail the second.

[14] As this chapter does, Ahmed (2013) distinguishes causal dependence and counterfactual dependence and notes that causal decision theory focuses on counterfactual dependence. However, he criticizes causal decision theory for focusing on counterfactual dependence instead of causal dependence. Because this chapter assumes causal decision theory, it saves for another occasion a response to his criticism.

[15] This section uses causal models to elucidate similarity among worlds. It does not adopt causal-model semantics for counterfactual conditionals. Briggs (2012) shows that a plausible version of causal-model semantics invalidates *modus ponens*, whereas adopting similarity semantics preserves that rule of inference.

Gibbard and Harper compute an option o's expected utility using for each state s the probability of the subjunctive conditional $P(o \,\square\!\!\rightarrow s)$ instead of the standard conditional probability $P(s \mid o)$. Causal relations affect the truth-conditions of a subjunctive conditional by affecting similarity relations among worlds. Section 2.3 replaces $P(o \,\square\!\!\rightarrow s)$ with $P(s$ given $o)$, a causal conditional-probability using subjunctive supposition of the condition. Subjunctive supposition of the condition o respects causal relations, as does subjunctive supposition of o in the conditional $o \,\square\!\!\rightarrow s$. Subjunctive supposition of the condition uses the same relations of similarity among worlds that subjunctive supposition of the conditional's antecedent uses. $P(s$ given $o)$ agrees with $P(o \,\square\!\!\rightarrow s)$ when the probability of the conditional exists. Given that for each option a unique nearest world realizing the option exists, $P(o \,\square\!\!\rightarrow s)$ equals $P(s$ given $o)$. The expected-utility formula's adoption of $P(s$ given $o)$ anticipates generalization to cases that do not meet the assumption of a unique nearest option-world. In cases in which the subjunctive conditional lacks a truth-value and its probability a value, the causal conditional-probability may nonetheless exist.

This section treats subjunctive conditionals such as $o \,\square\!\!\rightarrow s$ in cases in which a unique option-world exists and uses interventions in causal models to identify the nearest option-world. Its account of causal models comes from Pearl (2000) and Spirtes, Glymour, and Scheines (2000). A companion account of intervention in a causal model explains, given an option's supposition, which of the actual world's features stay fixed and which change to produce the nearest option-world.[16]

A causal model's representation is a directed acyclic graph (DAG) with variables serving as nodes. For convenience, one often says that two variables are causally related instead of saying that their values are causally related. In the case of a quantitative variable such as temperature, a value of the variable is not a number but an object's feature that the number represents and that may be a cause or an effect. Directed edges or arrows in the graph represent one variable's direct causal influence on another. This relation holds between variables an arrow connects if some values of the variable at the tail directly affect the value of the variable at the head. An arrow may represent a probabilistic causal influence between two variables. The model need not claim that if one variable has a certain value, the other variable has a certain value. It may yield a conclusion about an event's

---

[16] Pearl (2000: sec. 1.4.4); Spirtes, Glymour, and Scheines (2000: sec. 3.1); Spohn (2001); Hitchcock (2001); Woodward (2003: chap. 3); Sloman (2005: chap. 11); and Zhang (2012) use interventions in causal models to elucidate subjunctive conditionals.

Figure 6.2 A causal model

influence on another event's physical probability. A DAG is a causal graph if its edges represent direct causation. A causal graph represents causal relations among variables whose values influence the values of other variables. According to a causal model it depicts, one variable has an influence on another variable if a path of arrows goes from the first variable to the second.

An event-variable is a symbol, and its values may be event-tokens or event-types. A set of possible values represents the variable. The variable rain in a causal model's representation may have as values the event-type that it rains tomorrow and the event-type that it does not rain tomorrow. Suppose that a causal model has the variables R for rain's falling or not, S for the sprinkler's being on or off, and W for the grass's being wet or not. Figure 6.2 represents the model using a DAG.

A causal model for an agent's decision problem specifies causal relations among event variables, including an option variable that has as values the decision problem's options.[17] Indicator variables concerning realizations of BITs may be among the event variables that descend from the option variable. Setting the value of the option variable is an intervention. The agent intervenes to set the value of the option variable, and the model shows the value's influence on the values of event variables that descend from it. When setting the option variable's value, the intervention holds fixed the values of event variables that the option realized does not influence (the option variable's nondescendants). A deterministic causal model for a decision problem specifies what would happen if an intervention sets the option variable at a certain value (given values of nondescendant variables). It specifies the consequences of an option's realization (given values of nondescendant variables). An indeterministic model yields physical probabilities of consequences. Deliberations consider whether a BIT's realization is an option's consequence. If that cannot be discovered, then a fallback objective is to obtain a physical probability that the BIT's realization is the option's consequence. Evaluation of options using a causal model allows the option variable to take on its various values and derives an option's consequences or their physical probabilities from the causal model.

---

[17] Woodward (1997: 302) describes variables in causal models that decision theory uses.

Given complete information, an option's evaluation uses a single causal model to investigate the option's consequences. Consider the conditional that if a certain option were realized, then a certain avoidable realization of a BIT would occur. An intervention in the causal model, if the model is deterministic, settles the conditional's truth-value. It holds fixed all variables that are not descendants of the option variable and computes the option's consequences according to the model. If the BIT's indicator variable is a descendant of the option variable and giving the option variable a particular option as its value makes the BIT's indicator variable have the BIT's realization as its value, then the BIT's realization is a consequence of the option's realization. Because relations between variables may be probabilistic, setting the option-variable's value may boost the physical probability of a certain value of a descendant variable without making its physical probability equal one.

Given incomplete information, to obtain an option's expected utility, an evaluation begins with a set of mutually exclusive and jointly exhaustive causal models compatible with the agent's statistical data and background assumptions, given that the agent is certain of the data and assumptions. The evaluation assigns a probability to each compatible causal model. The effect of an intervention that sets the value of the option-variable depends not just on the causal model but also on the values of variables not descending from the option-variable (just as a cause's operation in a system depends on the system's initial conditions). An option's evaluation uses a causal model, along with values of variables not descending from the option variable, as a state of the world. The probabilities of causal models compatible with the evidence and the probabilities of values of the option variable's nondescendants generate the value of $P(o \,\square\!\!\rightarrow s)$ for an option o and state s.[18]

Pearl (2000: chap. 7) explains how intervention in a causal model for a decision problem settles the truth-values of subjunctive conditionals, such as those identifying an option's consequences. In general, an intervention given a causal model sets the value of a variable at a node without changing the causal relations governing other variables. Surgery eliminates every arrow leading to the node in a DAG representing the model as the intervention sets the value of the variable at the node. In a causal model for a decision problem, an option's realization intervenes to set the

---

[18] To gain generality, causal decision theory may replace $P(o \,\square\!\!\rightarrow s)$ with the probability image of s on o. Joyce (2010) uses causal models to ground probabilistic imaging.

value of the option variable while keeping fixed nondescendant variables. Interventions show what would happen if an option were realized.[19]

Pearl (2000: 108–10) criticizes evidential decision theory for conflating observation and intervention. In Pearl's terminology, an act, in contrast with an action, is an effect and an object of observation. An action, in contrast with an act, is an exercise of free will, a result of deliberation, and an intervention. Using his do-operator, Pearl represents an intervention by means of an action x as $do(x)$. He represents the content of an observation that the act x occurs as x without the do-operator. His do-calculus identifies the result of an intervention according to a causal model. To compute expected utilities, Pearl uses the probability of an outcome y given an intervention x, $P(y \mid do(x))$, whereas evidential decision theory uses the probability of the outcome y given the observation x, $P(y \mid x)$.[20] Evidential decision theory, Pearl claims, yields a mistaken recommendation in Newcomb's problem because it substitutes an act x for an action $do(x)$ in the conditional probabilities it uses to obtain expected utilities. It mistakenly uses an action as evidence for an outcome, as if it were an observation rather than an intervention.

Pearl's decision principle mishandles Newcomb's problem. Standard conditional probabilities take the condition $do(x)$ as evidence concerning other events and so lead expected-utility calculations to one-boxing. Contrary to Pearl, $P(C \mid do(A))$ does not equal $P(A \,\square\!\!\rightarrow C)$. The probability that the opaque box contains a million given that one picks only it does not equal the probability that it would contain a million if one were to pick only it. Also, the do-operation is not necessary for handling Newcomb's problem. After replacing $P(y \mid do(x))$ with $P(x \,\square\!\!\rightarrow y)$, calculations of expected utilities correctly recommend two-boxing in Newcomb's problem.

Consider the causal model for Newcomb's problem that Figure 6.3 presents. A choice is surgery on the arrow from choice disposition to choice. The subtraction sign before the arrow represents its cancellation. Given surgery, the choice is not evidence about winnings, according to Pearl. However, it is evidence about winnings because it is evidence about the prediction. $P(\text{one box predicted} \mid do(\text{take one box}))$ is high, and $P(\text{one box}$

---

[19] Equations in structural-equation causal models represent autonomous (invariant) mechanisms. Interventions knock out an equation. Pearl (2000: 204) says that interventions make minimal changes to causal mechanisms, or structural equations. Structural-equation models, given a world, specify the nearest worlds in which an intervention occurs.

[20] Freedman (1997: 120–21) suggests a distinction between two types of conditional probability. One is evidential and standard. The other is causal and registers the effect of an intervention, such as an intervention that makes a disease more or less probable. Freedman's distinction resembles the distinction that causal decision theory makes between evidential and causal conditional-probabilities.

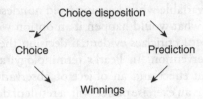

Figure 6.3 A causal model for Newcomb's problem

predicted | do(take two boxes)) is low, contrary to Pearl's claims. Pearl's account of surgery in the causal model for Newcomb's problem discards the evidence a choice provides about the prediction. Discarding that evidence is a mistake; an account of intervention with respect to a causal model should acknowledge that evidence.

Decision principles need not deny that a decision (and its execution) may provide evidence bearing on the outcome of the decision. In fact, decision principles must take account of such evidence to handle cases in which a person's choice supplies evidence about the outcome of her choice, for example, the cases Egan (2007) presents. Versions of causal decision theory that make a place for an option's self-ratification, as in Weirich (2001), do this.

Pearl (2000: 108) argues against probabilities of actions (that propositions represent) and treating actions as evidence. *Pace* Pearl, actions are not interventions from outside nature but are parts of nature. Consequently, probabilities attach to actions, and actions provide information about the world. Naturalism about actions acknowledges that a choice may offer evidence concerning its outcome. A choice's being free is compatible with its being evidence of the state of the world that settles the choice's outcome. Naturalistic decision principles acknowledge that a choice may be relevant evidence about its outcome.[21]

6.6 Causal inference

Probability assignments to an option's possible consequences rest on probability assignments to causal relations. This section briefly reviews principles

---

[21] Although Pearl's treatment of Newcomb's problem asserts that actions do not have probabilities, he assumes (2000: 108–10, 181) the existence of probabilities conditional on actions. Because the standard conditional probability $P(C \mid do(A))$ is undefined if $P(do(A))$ lacks a value, the existence of probabilities conditional on actions implies the existence of probabilities of actions. Allowing actions to have probabilities restores consistency.

of causal reasoning that go from evidence to assignments of probabilities to causal models and then to assignments of probabilities to causal relations.

Inferential methods find single-case causation by studying repetitions. Conclusions about causal relations between event-tokens typically rest on conclusions about causal relations between event-types manifest in repeated instantiations of the event-types and expressed in causal laws. Although causation among event-tokens is physically basic, beliefs about causal relations among event-types commonly grounds beliefs about causal relations among event-tokens.

In typical cases, causal inference moves from statistical data to conclusions about causal relations. It is inductive, not deductive, inference in these applications and generally yields conclusions with probability, not certainty. Reasonable inferences in some cases fail to arrive at true conclusions.[22]

Inductive reasoning from empirical data generates probability assignments to causal models involving variables with event-types as values. The assignments typically use methods of causal reasoning that incorporate assumptions. The probabilities of the assumptions that the methods employ affect the probabilities they yield. For instance, an assumption may state that an intervention does not destabilize the system a causal model represents. The assumption's probability affects the probability of a conclusion that the assumption grounds.

Epidemiology applies Mill's methods of causal inference. Suppose that all and only those who ate spinach (S) became ill (I). For the group of diners, a perfect correlation exists between the indicator variables S and I. Because of the correlation between S and I, a common assumption holds that either S influences I, I influences S, or some factor influences both S and I. The assumption does not rule out a common cause of spinach consumption and illness, such as the spinach server's being ill and distributing both spinach and illness. It does not rule out illness's causing spinach consumption by producing a taste for spinach. However, temporal order and background information, such as knowledge that unsanitary preparation of spinach may cause illness, strengthen the case for the spinach's causing the illness and weaken the case for rival causal relations. The correlation supports the

---

[22] Hanushek and Jackson (1977: chap. 1) take regression equations as linear causal models. They use samples to estimate the parameters, or variables' coefficients in the equation $Y = B_1 + B_2 X + U$. The variable Y may represent *Demand*, and the variable X may represent *Price*, for example. The equation may be part of a causal model, one given with structural equations. Freedman (1997) argues that regression models do not warrant deductive inferences moving from patterns of association to causal relations and that the assumptions on which inductive inferences to causal relations rest lack empirical confirmation in many cases.

inductive inference that S influences I, and background information strengthens the inference. Correlation does not warrant deductive conclusions about causation, but it grounds inductive conclusions.

The principle of the common cause, which Arntzenius (1992) and Papineau (1992) review, states that if two variables are correlated but neither causes the other, then they have a common cause. Sober (1988) argues that counterexamples exist: for a range of days, the price of bread in London is correlated with the sea level in Venice, although these variables do not have a common cause. Defenders of the principle look for a hidden common cause or hold that the principle demands a permanent correlation persisting in the future and not just a correlation that obtains up to the present. They may also observe that the principle expresses an inductive inference and tolerates exceptions.

In medicine, the double-blind controlled experiment sets the stage for an inference about a drug's efficacy. In the social sciences, controlled experiments often are not possible. Methods of causal inference use statistical data and assumptions about causation to eliminate some causal models involving the variables of interest. The methods improve Mill's by not requiring perfect correlation, and by explicitly dealing with rival causal models. Given assumptions about causation, they use statistical data to rule out some causal models. The principle of the common cause, for example, narrows the set of viable causal models. Background information such as information about temporal order supplements the methods.[23]

Evaluation of a causal model rests on statistical data about correlations between variables in the model and assumptions about causation. The evaluation yields a probability for a causal model that fits the data. Common methods of causal inference leave to judgment the assignment of a probability to a causal model. The methods of inference narrow the field of viable causal models and thereby assist Bayesian updating of a model's probability in light of new data. Some assumptions of the methods are built into the causal models that are candidates for an account of the causal relations among variables in the models.

An inductive inference from correlations that statistical data reveal to causation depends on background assumptions about plausible causal models.[24] A common assumption about viable causal models is the

---

[23] Gibbard (2008) explains how Bayesian methods use controlled experiments to update probabilities of causal relations.

[24] Pearl (2000: 49) claims that given a stability assumption, every joint probability distribution has a unique minimal causal structure (represented by a directed acyclic graph unique up to d-separation equivalence), as long as there are no hidden variables.

Causal Markov Condition. It states the results of screening off causes. Direct causes screen off indirect causes by making effects statistically independent of indirect causes given the direct causes. Pearl (2000: 61) states the Causal Markov Condition for a causal model this way: conditional on its parents, each variable is independent of its nondescendants. Causal inference eliminates causal models that violate the Causal Markov Condition in light of correlations that obtain after fixing the values of selected variables. Because exact conditional independence generally lacks empirical support, the method of elimination may use approximate conditional independence in the data to infer inductively conditional independence.[25]

Another assumption guiding elimination of causal models is Faithfulness. Faithfulness asserts that if variables are statistically independent, then neither causes the other. The Causal Markov Condition grounds inferences from causation to statistical data, and Faithfulness grounds inferences from statistical data to causation. Faithfulness is a simplicity condition; it eliminates causation not necessary for explaining the statistical data. It eliminates some causal models that are Markov-compatible with available statistical data.[26]

A Bayesian net represents a causal model. A Bayesian net is a DAG, together with a joint distribution of physical probabilities for its variables, provided that the combination satisfies the Causal Markov Condition and a simplicity condition such as Faithfulness.[27] Given a joint probability distribution for its variables, a DAG satisfies the Causal Markov Condition if and only if the joint probability distribution decomposes into a product of

---

[25] Some treatments of causal reasoning assume determinism so that probabilities in a causal model arise only from error terms associated with variables that have no parents. Provided that the error terms are independent, the model satisfies the Causal Markov Condition. This section does not assume determinism and uses the Causal Markov Condition to screen causal models. It uses the assumption to put aside indeterministic models in which two variables have correlated values because of a common parent, even after conditioning on the parent, as when the same mechanism flips two coins. The assumption is plausible in many cases of causal reasoning.

[26] Critics of the assumptions of causal modeling often target Faithfulness. They note that causal relations may cancel as in the case of birth control pills that increase the chance of thrombosis but also reduce the chance of pregnancy, which increases the chance of thrombosis. The pills may have no net effect on the chance of thrombosis but still are a cause of thrombosis. Such cases of perfect cancellation are rare. Usually, absence of correlation arises because of absence of causation. In fact, in cases in which empirical evidence supports a uniform probability distribution over possible causal models, a nonfaithful model has probability zero.

[27] According to one simplicity condition, the Minimality Condition, no proper subgraph satisfies the Causal Markov Condition. Zhang (2013) observes that simplicity motivates both the Minimality Condition and Faithfulness, and he distinguishes two forms of the Minimality Condition: Pearl's and also Sprites et al.'s.

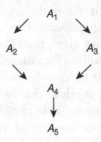

Figure 6.4 Decomposition of joint probabilities

standard conditional probabilities. Take, for example, the DAG in Figure 6.4. Let lowercase letters specify values of corresponding uppercase variables. Given that the model the figure represents satisfies the Causal Markov Condition, $P(a_1, \ldots, a_5) = P(a_1) \bullet P(a_2 \mid a_1) \bullet P(a_3 \mid a_1) \bullet P(a_4 \mid a_2, a_3) \bullet P(a_5 \mid a_4)$. Each conditional probability is the probability that a variable has a certain value given that the variable's parents have certain values.

Reasoning that supports a causal model rests on statistical data and assumptions concerning plausible causal models. An investigation begins with a set of models for variables of interest. The method of inference assumes that all viable causal models meet the Causal Markov Condition. Statistical data settle whether a model meets the condition. The data and assumptions eliminate models that fail to meet the condition. Considerations of simplicity, such as Faithfulness, eliminate other models. Using an example from Pearl (2000: 57), let R indicate the presence or absence of rain, S indicate whether the grass's sprinkling system is on, and W indicate the presence or absence of wet grass. Sprinkling makes grass wet without rain. So W does not cause R. Rain causes the grass to become wet, and not the other way around, because the sprinkler controls the grass's being wet without affecting rain. According to statistical data, R and W are probabilistically dependent, W and S are probabilistically dependent, and R and S are probabilistically independent. Causal reasoning, using the data and assumptions about plausible causal models, infers the model $R \rightarrow W \leftarrow S$ that Figure 6.2 presented.

For another example, consider the correlation between fire (F) and smoke (S). Does fire cause smoke? A controlled experiment may check whether manipulating smoke manipulates fire. If not, smoke does not cause fire. If controlled experiments are impossible, investigations may use naturally occurring correlations. They may infer correlations from sampling and statistical estimation of a joint distribution for physical probabilities. Then, they may apply the common-cause principle and assume that correlations arise from causation. According to the principle, F influences S, S influences F, or C influences both F and S for some variable C. To discover which alternative holds, they introduce a third variable, lightning (L), as a possible common cause. F's not being probabilistically independent of S given L eliminates the hypothesis that L directly influences both F and S, because according to that model and the Causal Markov Condition, F is probabilistically independent from S given L. Also gone are Markov equivalent models: $F \rightarrow L \rightarrow S$ and $S \rightarrow L \rightarrow F$. Then, L's being probabilistically independent of S given F eliminates the model $L \rightarrow F \leftarrow S$.[28] The model $L \rightarrow F \rightarrow S$ is compatible with S's being probabilistically independent of L given F. In fact, the Causal Markov Condition implies this probabilistic independence. If F and S are probabilistically independent given L, then L is a parent of one or both of F and S. This follows from Faithfulness in one form: all probabilistic independencies are due to the Markov condition. $L \rightarrow F \rightarrow S$ makes the probabilistic independence Markovian. Causal reasoning settles on the causal model $L \rightarrow F \rightarrow S$, or a class of models in which $F \rightarrow S$, and so concludes that $F \rightarrow S$. This conclusion follows deductively from the data and the assumptions. The assumptions such as Faithfulness and the correlations the assumptions use have inductive support, so the conclusion follows inductively from the data.

A decision problem generates a set of causal models with a variable having options as values and other descendant variables having possible consequences as values. If the option variable's values fall into classes of relevantly similar options, then an agent may have useful statistical data concerning relations between the option variable and other variables in the causal models. For example, the options may include smoking, and the agent may have statistical data about smoking's consequences for health. Methods of causal reasoning use available statistical data to identify causal models meeting the Causal Markov Condition and Faithfulness and to form a probability distribution over these models. The methods use the probability distribution to obtain the probabilities of causal relations between variables in the models and then between values of the option variable and descendant variables. Finally, for expected utility calculations, the methods obtain the probabilities of an option's possible consequences. This sketch omits

---

[28] In the eliminated model, F is a collider. The principle of d-separation, which follows from the Causal Markov Condition and a simplicity condition, states that conditioning on a collider makes the collider's parents probabilistically dependent.

many details but indicates how methods of causal reasoning supplement causal decision theory.

6.7   Causal utility's definition

This section introduces a type of utility, *causal utility*, that evaluates a proposition according to the proposition's consequences. Causal utility, using Section 6.4's account of an option's consequences, evaluates an option by its consequences exclusively.

Causal utility directly attends to consequences instead of locations of consequences, such as temporal intervals or spatiotemporal regions. In contrast with temporal utility and spatiotemporal utility, it evaluates BITs' realizations rather than locations hosting BITs' realizations. Temporal and spatiotemporal utility put aside many events certain not to be consequences (assuming that an option's consequences are in its future and in its light cone). Causal utility puts aside all such events (without making any empirical assumptions itself). A proposition's causal utility evaluates just the proposition's consequences. Among the BITs that would be realized if a proposition were realized, it attends to only those whose realizations are the proposition's consequences.

An option's consequences are events that would occur if the option were realized but would not occur if, for at least one other option, the other option were realized. Some events in an option's light cone would occur given each option's realization. They are unavoidable events and are not the option's consequences. An option's causal utility considers whether a BIT would be realized if the option were realized and would not be realized if some other option were realized. The causal relation it uses differs from other types of causal relation. An option's consequence, defined in terms of the causal relation, need not be an effect of the option. An option's consequences may extend farther than or not as far as do its effects. Causal utility's evaluation of an option asks for the option's consequences and only secondarily for the option's effects. If an option's consequences and effects come apart, its evaluation follows the option's consequences.

If realization of BIT_1 and realization of both BIT_1 and BIT_2 are consequences of o's realization, then adding the intrinsic utilities of both events double counts BIT_1's realization. To prevent double counting, an option's causal utility sums only the intrinsic utilities of basic consequences, namely, realizations of BITs. It puts aside realization of a pair of BITs. Causal utility evaluates all of an option's consequences, but its evaluation of them reduces to an evaluation of consequences that are realizations of BITs. A fully rational ideal agent assigns the same causal utility to an option evaluating either all its consequences or just those that are realizations of BITs. Because causal utility's evaluation of an option's consequences and its evaluation of the option's basic consequences are equivalent, applications may ignore the distinction.

Suppose that an option prevents realization of a BIT, and another option realizes the BIT. The prevention is the first option's consequence because it would occur if the option were realized and would not occur if some other option were realized.[29] Does causal utility take account of all consequences, including preventions, by evaluating only realizations of BITs?

In a decision problem, attention to preventions is not necessary for comparisons of options. Comparisons need not consider an option's prevention of a BIT's realization. They may concentrate on realizations of BITs. If an option prevents a BIT's realization, then the BIT's realization is a consequence of some other option. Causal utility's ranking of options counts an option's prevention of a BIT's realization by counting some other option's realization of the BIT. Instead of affecting the first option's evaluation, it affects the second option's evaluation and so factors into comparison of the two options. Causal utility, by attending for each option to only BITs whose realizations are the option's consequences, ranks options according to their choiceworthiness.

Suppose that one option has as a consequence a BIT's realization, and another option prevents its realization. Causal utility's comparison of the two options double counts their difference regarding the BIT if the causal utility of the option that realizes the BIT counts its realization, and the causal utility of the option that prevents the BIT's realization counts its prevention. Causal utility's omission of a BIT's prevention does not bias its ranking of options but instead eliminates redundancy. Section 6.8 verifies causal utility's ranking of options by showing that it is the same as world-Bayesianism's ranking.[30]

---

[29] In a decision problem, for an option o and a state s, if s is a consequence of o, then, according to the definition of a consequence, $(o \,\square\!\!\rightarrow s)$ and $\sim(o' \,\square\!\!\rightarrow s)$ for some option o'. The law of conditional excluded middle for conditionals having options as antecedents follows from the chapter's simplifying assumption that for each option, a nearest option-world exists. From $(o \,\square\!\!\rightarrow s)$, it follows that $\sim(o \,\square\!\!\rightarrow \sim s)$, and, given conditional excluded middle, from $\sim(o' \,\square\!\!\rightarrow s)$, it follows that $(o' \,\square\!\!\rightarrow \sim s)$. Consequently, $(o' \,\square\!\!\rightarrow \sim s)$ and $\sim(o \,\square\!\!\rightarrow \sim s)$. Thus, by the definition of a consequence, $\sim s$ is a consequence of o'. In general, if s is a consequence of o, then $\sim s$ is a consequence of o' for some o'.

[30] Consider a rare case in which p and $\sim p$ are both objects of BITs. Suppose that an athlete has a basic intrinsic desire to win a match and also a basic intrinsic aversion toward losing the match. If comparison of options uses realizations of BITs, does it double count considerations when comparing an option that yields winning and another option that yields losing? No, because the desire to win and the aversion toward losing are independent, the options' comparison should use both BITs.

Given complete information, an option's causal utility is the sum of the intrinsic utilities of objects of BITs whose realizations are the option's consequences. Allowing for incomplete information, it is the probability-weighted average of this quantity in the worlds that might be the option's world.

The definition of causal utility for an arbitrary proposition is a straight-forward extension of the definition for an option. Given complete information, a proposition p's causal utility, CU, is the sum of the intrinsic utilities of the objects of BITs whose realizations are avoidable but would be realized if p were realized: $CU(p) = \sum_i IU(\text{BIT}_i)$, where BIT_i ranges over the objects of BITs whose realizations are p's consequences. Allowing for incomplete information, causal utility evaluates a proposition p by examining each possible world that realizes p. In a world, consequences are events avoidable in the world. For each p-world, causal utility attends to BITs the world realizes but whose realizations are avoidable in that world. Thus, the definition of p's causal utility is

$$CU(p) = \sum_i P(w_i \text{ given } p) \sum_j IU(\text{BIT}_{ij})$$

where w_i ranges over the worlds that might be realized given p, and BIT_{ij} ranges over the objects of BITs w_i realizes but whose realizations are avoidable in w_i.[31]

For an illustration of the definition, take the causal utility of playing cricket assuming that the only relevant possible consequences are realizations of basic intrinsic desires for fun and for health. Assume that fun comes for sure and health with probability ½. The two relevant possible worlds offer, first, fun and health and, second, just fun. Then, $CU(\text{cricket}) = (½)[IU(\text{fun}) + IU(\text{health})] + (½)IU(\text{fun})$.

The definition of conditional causal utilities, factors in expected causal utilities, is a straightforward extension of the definition of causal utilities. Given complete information, a proposition's conditional causal utility is a sum of the intrinsic utilities of the proposition's basic consequences given the condition: $CU(p \text{ given } q) = \sum_i IU(\text{BIT}_i)$, where BIT_i ranges over the objects of BITs whose realizations are avoidable but would be realized, given q, if p were realized. Allowing for incomplete information, conditional causal utility evaluates a proposition by evaluating each proposition-world meeting the condition. For each such proposition-world, it puts aside events

---

[31] An alternative definition says that a proposition's causal utility evaluates just the proposition's consequences. According to it, the proposition's causal utility equals the probability-weighted sum because of a norm for ideal agents rather than because of a definition.

that are unavoidable in the world and considers the proposition's basic consequences in the world:

$$CU(p \text{ given } q) = \sum_i P(w_i \text{ given } p)\sum_j IU(\text{BIT}_{ij})$$

where w_i ranges over the worlds that, given q, might be realized given p, and BIT_{ij} ranges over the objects of BITs whose realizations are avoidable but would be realized in w_i. The chapter's appendix presents methods of calculating causal utilities, including applications of expected-utility analysis, and demonstrates the consistency of various methods of calculating causal utilities.

6.8 Causal-utility analysis

Causal-utility analysis breaks a causal utility into components. A form of causal-utility analysis divides the causal utility of an option's world into (1) the option's causal utility, an evaluation of the option's consequences, and (2) an evaluation of events besides the option's consequences. It sorts realizations of BITs in an option's world into those that are and those that are not the option's consequences. The sorting grounds the separability of an option's consequences from other events, as Chapter 3's general principles maintain. This section presents only the analysis just sketched of the causal utility of an option's world. Causal utility's definition accommodates evaluation of all propositions, but this section does not present a general form of causal-utility analysis for all propositions and all ways of dividing a proposition's causal utility into parts. The chapter's appendix presents a versatile form of causal-utility analysis that calculates an option's causal utility directly from the intrinsic utilities of the BITs the option might realize.

In a decision problem, this section's form of causal-utility analysis divides an option's world, not into two regions, but into two classes of events: the option's consequences and other events. Because causal utility evaluates directly a class of events rather than a region, a ranking of options may apply causal utility directly to options. Causal utility limits the scope of its evaluations of options without introducing regions to evaluate. Although the union of the regions of an option's basic consequences forms an appropriate region to evaluate, causal utility need not evaluate an option's consequences by evaluating their region. It identifies an option's consequences independently of their locations in a region.

Temporal-, spatiotemporal-, and causal-utility analyses all use a single type of utility to evaluate an option's world and the option itself; for example, causal-utility analyses use causal utility for both. Temporal- and spatiotemporal-utility analyses also use a single type of utility to evaluate an option and its background in a world; for example, temporal-utility analyses use temporal utility for both. However, causal-utility analyses differ in this respect. The causal utility of a world equals its intrinsic utility (assuming that BITs attach only to contingent events), whereas the causal utility of an option's nonconsequences equals, instead of their intrinsic utility, the intrinsic utility of the option's world.[32] So a causal-utility analysis does not divide a causal utility into two causal utilities but into one causal utility and an intrinsic utility. It uses an option's causal utility and its nonconsequences' intrinsic utility. Support for the analysis considers first its application to cases of complete information and then its application to cases allowing for incomplete information.

Given complete information, causal-utility analysis treats only the option's world. The relevant events in the world are its realizations of BITs. The analysis divides the BITs the world realizes into BITs whose realizations are the option's consequences and the rest. Each BIT falls into exactly one class. So, the world's intrinsic utility is the sum of the intrinsic utilities of its parts, taking each class of BITs to be represented by the conjunction of the objects of the BITs in the class. If a conjunction of objects of BITs entails realization of a BIT, the BIT's object is a conjunct. A contingent event that an option's consequences entail is the option's consequence. An event that an option's nonconsequences entail is the option's nonconsequence. An option's causal utility, as Section 6.7 defines it, is a sum of intrinsic utilities of objects of BITs whose realizations are the option's consequences. This sum plus the intrinsic utility of the conjunction of objects of BITs whose realizations are unavoidable in the decision problem yields the intrinsic utility of the option's world. Section 3.4 shows that the comprehensive utility of an option's world equals the world's intrinsic utility. A causal-utility analysis shows that the intrinsic utility of an option's world equals the sum of the option's causal utility (namely, the intrinsic utility of the option's consequences in the world) and the intrinsic utility of the option's nonconsequences in the world.

[32] A world's causal utility evaluates consequences of the world's realization. An event in the world is not a consequence unless it fails to occur in some alternative to the world. Context makes every other world an alternative. So, every contingent event in the world is a consequence of the world's realization. Assuming that BITs attach only to contingent events, the world's causal utility equals its intrinsic utility; all BITs a world realizes are avoidable in it because the relevant alternatives to the world include worlds that do not realize the BITs. If some BITs attach to necessary events, the causal utility of an option's world may not equal the world's intrinsic utility, but the world's intrinsic utility still divides into the option's causal utility and the intrinsic utility of the option's nonconsequences in the world.

An option's nonconsequences are the same for each option. Also, an option's comprehensive utility equals its world's intrinsic utility. An option's causal utility therefore equals its comprehensive utility except for subtraction of a constant that is the same for all options. Thus, causal utility and comprehensive utility rank options the same way. Comprehensive utility's order of options according to their worlds is the same as causal utility's order of options according to their consequences. Hence, an evaluation of options may use causal utility instead of comprehensive utility. Using options' causal utilities yields the same comparisons of options as using options' comprehensive utilities.

In the general case, allowing for incomplete information, an option's comprehensive utility equals the expected value of its world's intrinsic utility. For each world that might be the option's world, causal-utility analysis divides the world's realizations of BITs into the option's consequences and the rest. The sum of the intrinsic utilities of the two classes equals the world's intrinsic utility. The expected value of the intrinsic utility of the class of consequences of the option equals the option's causal utility, as Section 6.7 defines it. The events that are not an option's consequences in a world that might be the option's world are the unavoidable events in the world. The unavoidable events, although they may be unknown, are the same for all options. The class of BITs whose realizations are not consequences is the same for all options, and its expected intrinsic utility is the same for all options. Because comprehensive utility, the expected intrinsic utility of an option's world, evaluates the worlds that might be an option's world, and causal utility evaluates just the option's consequences in those worlds, $U(o) = CU(o) + \sum_i P(w_i \text{ given } o) \sum_j IU(\text{BIT}_{ij})$, where w_i ranges over the worlds that might be realized given o, and BIT_{ij} ranges over the objects of BITs w_i realizes and whose realizations are unavoidable and so fail to be o's consequences. As explained, the formula's last factor, the expected intrinsic utility of o's nonconsequences, is the same for every option o. Hence, $U(o)$ equals $CU(o)$ after adding a constant to the latter. So, the options' causal utilities and their comprehensive utilities rank options the same way. This follows from Section 3.7's general principle of separability.

Causal utility not only ranks options as comprehensive utility does but also ranks options efficiently because it processes only relevant considerations, namely, avoidable realizations of BITs, or basic consequences. The efficiency in evaluation of options that comes from processing only relevant considerations is just one type of efficiency. Using pair-wise comparisons of options also improves causal utility's overall efficiency in ranking options. Moreover, gaining efficiency in calculating causal utilities, in comparing options with respect to causal utility, and in executing the steps of a ranking of options according to causal utility improves causal utility's overall efficiency in ranking options. Nonetheless, using causal utilities to evaluate options is efficient with respect to considerations the evaluations entertain. In this respect, evaluation by causal utility is a maximally efficient general method of ranking options the same way that comprehensive utility ranks them; all other general methods process irrelevant considerations.

6.9 Generality

Evaluation of options, for unity, needs a single method of expected-utility analysis that, for any partition of states, applies to any type of utility that ranks options. Section 2.5's formulation of expected-utility analysis anticipates its extension to causal utility. This section explains its extension.

In cases in which options have no causal influence on states, some versions of expected-utility analysis advance the equation $U(o) = \sum_i P(s_i) U(o \& s_i)$, according to which the utility of an option-state pair is the utility of the conjunction of the option and state. The analog of this equation for causal utility is $CU(o) = \sum_i P(s_i) CU(o \& s_i)$. This way of applying expected-utility analysis to causal utility fails. $CU(o \& s_i)$ evaluates consequences of s_i as well as consequences of o, whereas $CU(o)$ evaluates only consequences of o. A revision of expected-utility analysis for causal utility replaces $CU(o \& s_i)$ with $CU(o$ given $s_i)$ as the utility for an option-state pair. That is, it uses a conditional causal utility instead of a causal utility of a conjunction. Plainly, the two causal utilities may differ. Whereas $CU(o \& s_i)$ equals $CU(s_i \& o)$, $CU(o$ given $s_i)$ differs from $CU(s_i$ given $o)$ when o's consequences given s_i are relevantly different from s_i's consequences given o. Although $CU(o \& s_i)$ appraises consequences of o and consequences of s_i, $CU(o$ given $s_i)$ appraises only consequences of o.

After substituting conditional utilities for utilities of conjunctions, expected-utility analysis advances the equation $CU(o) = \sum_i P(s_i) CU(o$ given $s_i)$ for cases in which states are causally independent of options. Generalizing to allow for cases in which an option has a causal influence on a state, expected-utility analysis makes probabilities of states conditional on the option. The result for causal utility states that $CU(o) = \sum_i P(s_i$ given $o) CU(o$ given $s_i)$, where the probabilities are causal conditional-probabilities, as Section 2.1 characterizes them.

The interpretation of $CU(o$ given $s_i)$ that best suits the formula for $CU(o)$ evaluates the consequences that o *would* have if o *were* realized given that s_i *is* the case. That is, it uses subjunctive supposition for o and indicative supposition for s_i. This interpretation of $CU(o$ given $s_i)$ allows expected causal utility to be defined given any partition of states. It is defined, for instance, given a partition that takes states as options. Given the difference in forms of supposition for option and state, $CU(o$ given $\sim o)$ is coherent. It is the causal utility of the consequences that o would have if o were realized given that $\sim o$ is in fact the case, that is, given that another option is realized. The interpretation of $CU(o$ given $s_i)$ thus ensures that expected-utility analysis applied to causal utility may use any partition of states.

This interpretation of $CU(o$ given $s_i)$, despite its virtues, needs refinement to make $CU(o)$ equal $\sum_i P(s_i$ given $o)CU(o$ given $s_i)$. Given some ways of supposing s_i, $CU(o$ given $s_i)$ does not entertain the possibility that s_i is a consequence of o. Suppose that s_i is avoidable, and if o were realized s_i would obtain, so that s_i is a consequence of o. $CU(o$ given $s_i)$ ignores s_i's being a consequence of o if its supposition that s_i is the case fixes s_i, makes it unavoidable, and so not a consequence of o. Section 6.7 defines $CU(o)$ as the expected intrinsic utility of a full specification of o's consequences according to o's world. Because s_i is a consequence of o, a computation of $CU(o)$ should include this consequence among o's consequences, and so $CU(o$ given $s_i)$ should do the same. The conditional causal utility must not assume that s_i holds independently of the option realized.

Section 2.5's general formulation of expected-utility analysis uses for each option o and state s_i the conditional utility $U(o$ given $s_i)$, interpreted so that it equals $U(o$ given $(o \,\square\!\!\rightarrow s_i))$. This interpretation of conditional utility grounds expected-utility analysis's extension to causal utility. To count all relevant consequences, a calculation of $CU(o)$ using expected-utility analysis interprets $CU(o$ given $s_i)$ so that it agrees with $CU(o$ given $(o \,\square\!\!\rightarrow s_i))$. That is, it takes o's consequences given that s_i to be the same as o's consequences given that it *is* the case that s_i *would* obtain if o *were* performed. The complex condition is in the indicative mood but concerns a conditional in the subjunctive mood. $CU(o$ given $(o \,\square\!\!\rightarrow s_i))$ evaluates an option o under indicative supposition of the subjunctive conditional $(o \,\square\!\!\rightarrow s_i)$, assuming that a unique nearest o-world exists. The evaluation reviews the BITs whose realizations might be o's consequences under indicative supposition of the subjunctive conditional. Applying the subjunctive mood to s_i allows it to be entertained as a consequence of o. The nuanced interpretation of $CU(o$ given $s_i)$ furnishes a satisfactory account of an option-state pair's causal utility.[33]

---

[33] Which mood does a statement of the conditional causal utility use to state the condition? In "$CU(o$ given $s_i)$", "given s_i" is short for "given s_i's truth". This way of stating the condition does not adopt a

Section 2.5's case concerning the helpfulness of rising to close a door illustrates the interpretation's success. In this case, whether John is helpful depends on whether a door closes as a result of John's rising to close it. Given indicative supposition of s_i, $CU(o$ given $s_i)$ does not allow s_i to count as a consequence of o. So, the causal utility of rising given the door's closing does not take account of the possibility that the door's closing is a consequence of rising. On the other hand, interpreting $CU(o$ given $s_i)$ to agree with $CU(o$ given $(o \,\square\!\!\rightarrow s_i))$ allows s_i to count as a consequence of o. It uses a type of supposition of s_i that does not assume s_i holds independently of o. So the causal utility of rising's consequences, given that the door would close if John were to rise to close it, takes account of the possibility that John closes the door. It generates the right probability-utility product for an option-state pair in an expected-utility analysis of rising's causal utility.

$CU(o$ given $s_i)$ evaluates the consequences of o given s_i. The consequences of o come from an intervention's setting the option variable at o. The consequences may be probabilistic within a causal model, and then $CU(o$ given $s_i)$ is a probability-weighted average of the utilities of the consequences of o given s_i, that is, their expected utility. $CU(o$ given $s_i)$ equals $CU(o$ given $(o \,\square\!\!\rightarrow s_i))$ if the second conditional utility exists. $CU(o$ given $(o \,\square\!\!\rightarrow s_i))$ exists under the assumption that $(o \,\square\!\!\rightarrow s_i)$ is a proposition with a truth-value. If $(o \,\square\!\!\rightarrow s_i)$ does not have a truth-value, then, although $CU(o$ given $(o \,\square\!\!\rightarrow s_i))$ does not exist because its supposition lacks suitable content, $CU(o$ given $s_i)$ exists.

Adopting this section's interpretation of $CU(o$ given $s_i)$, the general equation for expected causal-utility is

$$CU(o) = \sum\nolimits_i P(s_i \text{ given } o)\,CU(o \text{ given } s_i)$$

The chapter's appendix shows that this form of expected causal-utility analysis agrees with Section 6.7's definition of causal utility. Adopting it unifies methods of evaluating options.

This chapter introduced causal utility and showed how to use it to evaluate options according to their consequences. It defined consequences using conditionals and considered how causal models may guide assessments of conditionals and guide assignments of probabilities to possible consequences of options. Causal utility yields the most efficient general method of evaluating options. It simplifies choices by processing only relevant considerations.

mood for its statement, as Section 2.6 observes. The theory of conditional causal utility specifies a type of supposition for the condition. The remarks on future contingencies in Weirich (1980), discussed by Davis (1982), consider how $CU(o$ given $s_i)$ settles the form of supposition for s_i.

6.10   **Appendix: Calculation and consistency**

This appendix presents a method of calculating an option's causal utility, shows that the definition of causal utility agrees with expected-utility analysis, and shows that the definition of causal utility agrees with the canonical form of utility analysis.

Calculation of an option's causal utility may consider whether, for each BIT, the BIT's realization is the option's consequence instead of (1) introducing possible states (such as causal models and values for the option-variable's nondescendants) and (2) identifying in each state the BITs realized or their probabilities of realization. States are intermediaries between options and consequences, and causal-utility analysis may skip the intermediaries. An option's evaluation need not ask whether, if a state holds, a BIT's realization is an option's consequence. It may ask directly whether the BIT's realization is the option's consequence. If the answer is uncertain, it may use the probability that the BIT's realization is the option's consequence. Judgment assisted by reasoning about causal models assigns the probability, so the direct question about a BIT's realization does not ignore states but just puts them in the background.

The direct method of calculating an option's causal utility considers for each BIT whether its realization is among the option's consequences in a world that might be the option's world. It sums utilities of (epistemic) chances that objects of BITs are the option's consequences. The utility of a chance that a BIT's realization is o's consequence, because it is o's consequence in a particular possible world, equals the probability of the world times the intrinsic utility of the BIT's realization. The calculation may aggregate the probabilities of worlds in which a BIT's realization is o's consequence. The probability that a BIT's realization is o's consequence by definition is the probability that $(o \ \Box\!\!\rightarrow BIT) \ \& \ \exists o' \sim (o' \ \Box\!\!\rightarrow BIT)$, letting BIT stand for the BIT's object. This probability equals the sum of the probabilities of the worlds in which the BIT's realization is o's consequence, that is, the worlds in which the conjunction is true. An option o's causal utility thus equals a probability-weighted sum of the intrinsic utilities of the objects of BITs that might be the option's consequences: $CU(o) = \sum_i P(BIT_i$ is a consequence of $o)IU(BIT_i)$, where BIT_i ranges over all BITs' objects. Efficiency eliminates BITs whose realizations are certainly not consequences; the probability-utility products for them equal zero. In particular, it eliminates BITs that the agent is certain are realized unavoidably.

Next, consider the general form of expected causal-utility analysis. It may calculate an option's causal utility using states coarser than worlds, whereas the definition of causal utility uses worlds. An expected causal-utility analysis of $CU(o)$ and the definition of $CU(o)$ agree because both rely on evaluation by BITs. Because expected causal-utility analyses agree with each other, a case for their consistency with the definition need only show agreement between some expected causal-utility analysis and the definition. In an expected causal-utility analysis of $CU(o)$, let each state s_i be a possible full description of o's basic consequences in a world w_i that might be realized given o. Then, Section 6.9's causal-utility analysis, $CU(o) = \sum_i P(s_i \text{ given } o) CU(o \text{ given } s_i)$, implies that $CU(o) = \sum_i P(s_i \text{ given } o) CU(s_i)$ because s_i's basic consequences are the same as o's basic consequences given s_i. (Recall that a proposition's realization counts as its own consequence.) According to Section 6.7's definition of causal utility, $CU(o) = \sum_i P(w_i \text{ given } o)\sum_j IU(\text{BIT}_{ij})$, where w_i ranges over the worlds that might be realized given o, and BIT_{ij} ranges over the objects of BITs whose realizations are avoidable but would be realized in w_i, that is, o's basic consequences in w_i. It follows that $CU(o) = \sum_i P(s_i \text{ given } o) IU(s_i)$, where s_i ranges over full descriptions of o's basic consequences in the worlds that might be realized given o. The analysis agrees with the definition because $CU(s_i) = IU(s_i)$.

Last is a demonstration that in a decision problem the definition of causal utility is consistent with the canonical form of utility analysis, that is, the combination of expected- and intrinsic-utility analyses. Consider the causal utility CU of an option o's outcome $O[o]$, a possible world. All the world's contingent events are its consequences because each would occur given the world's realization and for some alternative world would not occur given the alternative's realization. Thus, $CU(O[o]) = U(O[o]) = U(o)$, given that no BITs attach to necessary events. For consistency, the definition of $CU(O[o])$ must agree with a canonical analysis of $U(o)$. The next two paragraphs establish their agreement.

According to Section 6.7's definition of causal utility, $CU(p) = \sum_i P(w_i \text{ given } p)\sum_j IU(\text{BIT}_{ij})$, where w_i ranges over the worlds that might be realized given p, and BIT_{ij} ranges over the objects of BITs whose realizations are avoidable but would be realized in w_i, that is, p's basic consequences in w_i. According to Section 3.4, a canonical analysis of $U(o)$ uses expected-utility analysis and intrinsic-utility analysis: $U(o) = \sum_i P(w_i \text{ given } o)\sum_{j \in \{ij\}} IU(\text{BIT}_j)$, where $\{ij\}$ indexes the BITs realized in w_i.

By the definition of causal utility, $CU(O[o]) = \sum_i P(w_i \text{ given } O[o])\sum_j IU(\text{BIT}_{ij})$, where w_i ranges over the worlds that might be realized given $O[o]$, and BIT_{ij} ranges over the objects of BITs whose realizations are avoidable but would be realized in w_i. The BITs' realizations are $O[o]$'s basic consequences in w_i. The range of BITs covers exactly the BITs w_i

realizes, given that no BITs attach to necessary events. $P(w_i$ given $O[o]) =$ $P(w_i$ given $o)$, and w_i might be realized given $O[o]$ if and only if w_i might be realized given o. So, $CU(O[o]) = \sum_i P(w_i$ given $o)\sum_j IU(\text{BIT}_{ij})$, where w_i ranges over the worlds that might be realized given o, and BIT_{ij} ranges over the objects of BITs that w_i realizes. $U(o)$ equals the sum on the right-hand side of the equation, by the canonical analysis of $U(o)$. Hence, the definition of $CU(O[o])$ is consistent with a canonical analysis of $U(o)$.

Conclusion

Efficiency tells deliberation to put aside irrelevant considerations. Value holism warns against ignoring any considerations; even factors of no value, because of complementarities with other factors, may affect an option's value, it observes. Efficiency gains the upper hand in this debate. Using intrinsic-utility analysis, deliberation can put aside irrelevant considerations without overlooking relevant complementarities.

This book formulates and justifies accurate general methods of evaluating the options in a decision problem after putting aside irrelevant considerations. Chapters 1–3 set the stage by defining separability of considerations and by establishing the separability of chances and the separability of realizations of basic intrinsic attitudes. Chapters 4–6 use separability of considerations to justify evaluation of options according to their futures, regions of influence, and consequences.

Intrinsic-utility analysis, which reasonable forms of operationalism permit, decomposes a world's intrinsic utility into the intrinsic utilities of the world's realizations of basic intrinsic attitudes (BITs). Given incomplete information about an option's world, an option's evaluation may separate chances for worlds and separate BITs that worlds realize. Intrinsic utility's separability is fundamental. Principles of intrinsic utility justify separating considerations to focus on relevant considerations. Given complete information, they sum intrinsic utilities of BITs' realizations to calculate the intrinsic utilities of an option's future, events in its region of influence, and its consequences. Given incomplete information, they combine with expected-utility analysis to compute estimates of these quantities. Intrinsic utility's separability grounds temporal, spatiotemporal, and causal utilities' separability. Their separability permits an option's evaluation to assess the option's future, region of influence, or consequences.

Chapter 1 presents the key concept of separability. Chapter 2 justifies expected-utility analysis's separation of an option's utility into utilities of chances for possible outcomes. The argument uses the independence of the chances' utilities given that the possible outcomes are comprehensive and thereby include interaction effects. Chapter 3 justifies intrinsic-utility analysis's separation of the utility of an option's world into utilities of the world's realizations of BITs. The argument uses the independence of the utilities of BITs' realizations. Combining intrinsic- and expected-utility analyses justifies the canonical form of separating considerations using an option's chances of realizing possible worlds and the worlds' intrinsic utilities.

Common forms of utility analysis evaluate an option using evaluations of temporal segments of the option's world, spatiotemporal segments of the option's world, and the option's consequences in its world. Chapters 4–6 make precise and justify these forms of utility analysis. The chapters use types of utility with restricted evaluative scope and specify assignments of events to an option's future, its region of influence, and its set of consequences. Because no BIT's realization straddles the break between a world part and its complement, the forms of utility analysis neither omit nor double count any relevant consideration. Each type of utility handles uncertainty by computing the expected value of its type of utility for an option. Distinct chapters treat the forms of utility analysis separately, despite their common structure, because each form's presentation requires specifying details, such as the type of causal relation that grounds causal-utility analysis.

A world's comprehensive utility equals the world's intrinsic utility, that is, the sum of the intrinsic utilities of the objects of BITs that the world realizes. In a decision problem, the options' worlds share realizations of many BITs. Subtracting the intrinsic utilities of objects of BITs that all worlds realize simplifies comparisons of options. Subtracting a common factor from a set of sums does not change the sums' order (because summation is a separable function). Subtracting intrinsic utilities of realizations of BITs in an option's past, or in the complement of the option's region of influence, or in the complement of the option's consequences simplifies the option's evaluation. A ranking of options after such a subtraction for each option agrees with the options' ranking according to ordinary comprehensive utilities and so may direct a decision.

The chapters' simplified methods of evaluating options introduce temporal utility, spatiotemporal utility, and causal utility as subutility functions with respect to the intrinsic-utility function for a world. Intrinsic-utility analysis establishes separability with respect to an all-factor utility function for a world, from which restricted utility functions arise by fixing some arguments of the all-factor function and letting only the remaining arguments vary. For example, the past is the same for all options, so the set of arguments representing the past in the all-factor function is the same for all options. Holding fixed these arguments yields a restricted function of the remaining arguments, in particular, those settling the future. The same holds for other ways of fixing argument places. Because of separability, the restricted utility functions rank options just as the all-factor function does.

Evaluation by consequences is a maximally efficient general way of ranking options as comprehensive utility ranks them. Although other evaluations of options are useful, appraisal of options using consequences processes the fewest considerations. It achieves a theoretical type of efficiency and also practical efficiency in cases in which obstacles to its application are minor because, for instance, each option's consequences are easy to identify. Imagine a decision problem for a corporation in which profits are the only relevant consequences, and its options (say, whether or not to build a new warehouse) have identifiable expected monetary consequences. An efficient evaluation of the corporation's options concentrates on each option's possible consequences. Their being monetary makes them easy to identify.

Temporal, spatiotemporal, and causal utilities have increasingly narrow evaluative scope. Temporal-, spatiotemporal-, and causal-utility analyses separate considerations so that an evaluation of options may examine options' futures, regions of influence, or consequences. Although evaluating options' consequences is in general most efficient theoretically, circumstances may make evaluating options' futures or regions of influence more practical.

Suppose that the federal government changes its tax laws. Ideally, an evaluation examines the consequences for each citizen. Given the difficulty of assessing these consequences, it may be more practical to evaluate changes in each of the fifty states, taking account of their various tax laws. If that analysis is too demanding, the most practical evaluation may attend to the nation's future under the new tax laws. The most practical evaluation is not in every case the general form of evaluation that is theoretically most efficient.

The usefulness of a normative model's efficient methods of evaluating options depends on how closely people resemble the model's ideal agents. Studies in managerial psychology assume that workers have intrinsic and extrinsic desires and recommend that a plan for compensating workers consider assigning tasks to workers to enhance intrinsic rewards. Whether workers' desires have a structure that BITs usefully represent is a topic for research in managerial psychology.

Future research in normative decision theory will further explain an option's consequences and devise methods of identifying them (and their probabilities). It will formulate methods of inductively inferring options' consequences (and their probabilities). A general account of an option's consequences will handle cases in which subjunctive conditionals do not adequately characterize the option's consequences.

Traditional efficiency-promoting methods of evaluating options are justifiable and valuable components of decision theory. This book's normative model for ideal agents constructs a congenial environment for these methods. Generalizing the model by removing its idealizations and adjusting the methods to handle obstacles will show how human deliberations may reliably use the methods. An accurate normative model using idealizations positions decision theory to move closer to a general, realistic account of efficient deliberations.

References

Ahmed, A. 2013. "Causal Decision Theory: A Counterexample." *Philosophical Review* 122: 289–306.

Arntzenius, F. 1992. "The Common Cause Principle." *PSA: Proceedings of the Biennial Meeting of the Philosophy of Science Association, Vol. 2, Symposia and Invited Papers*, pp. 227–37.

Baron, J. 2008. *Thinking and Deciding*. Fourth edition. Cambridge: Cambridge University Press.

Bentham, J. [1789] 1987. "An Introduction to the Principles of Morals and Legislation." In *Utilitarianism and Other Essays*. Harmondsworth, Middlesex, UK: Penguin Books.

Bernoulli, D. [1738] 1954. "Exposition of a New Theory on the Measurement of Risk." *Econometrica* 22: 23–36.

Binmore, K. 1998. *Game Theory and the Social Contract, Vol. 2, Just Playing*. Cambridge, MA: MIT Press.

2007. *Game Theory: A Very Short Introduction*. New York: Oxford University Press.

2009. *Rational Decisions*. Princeton, NJ: Princeton University Press.

Black, J. 2002. *A Dictionary of Economics*. Second edition. Oxford: Oxford University Press.

Blackorby, C., D. Primont, and R. Russell. 2008. "Separability." In S. Durlauf and L. Blume, eds., *New Palgrave Dictionary of Economics*, pp. 431–35. New York: Macmillan.

Bradley, R. 1999. "Conditional Desirability." *Theory and Decision* 47: 23–55.

2012. "Multidimensional Possible-World Semantics for Conditionals." *Philosophical Review* 121: 539–71.

Briggs, R. 2012. "Interventionist Counterfactuals." *Philosophical Studies* 160: 139–66.

Broome, J. 1991. *Weighing Goods*. Oxford: Blackwell.

Brown, C. 2007. "Two Kinds of Holism about Values." *Philosophical Quarterly* 57: 456–63.

Buchak, L. 2013. *Risk and Rationality*. Oxford: Oxford University Press.

Carlson, E. 2001. "Organic Unities, Non-Trade-Off, and the Additivity of Intrinsic Value." *Journal of Ethics* 5: 335–60.

Carnap, R. [1950] 1962. *Logical Foundations of Probability*. Second edition. Chicago: University of Chicago Press.

Chang, H. 2004. *Inventing Temperature: Measurement and Scientific Practice*. New York: Oxford University Press.

2009. "Operationalism." In E. Zalta, ed., *Stanford Encyclopedia of Philosophy*. http://plato.stanford.edu/archives/fall2009/entries/operationalism/

Chisholm, R. 1975. "The Intrinsic Value in Disjunctive States of Affairs." *Noûs* 9: 295–308.

1986. *Brentano and Intrinsic Value*. Cambridge: Cambridge University Press.

Christensen, D. 2004. *Putting Logic in Its Place*. Oxford: Clarendon Press.

Dancy, J. 1993. *Moral Reasons*. Oxford: Blackwell.

2004. *Ethics without Principles*. Oxford: Clarendon Press.

Davis, W. 1982. "Weirich on Conditional and Expected Utility." *Journal of Philosophy* 79: 342–50.

de Finetti, B. 1937. "La Prevision." *Annales de l'Institute Henri Poincaré* 7: 1–68.

Dietrich, F. and C. List. 2013. "Reason-Based Rationalization." Manuscript.

Dreier, J. 1996. "Rational Preference: Decision Theory as a Theory of Practical Rationality." *Theory and Decision* 40: 249–76.

Dretske, F. 1988. *Explaining Behavior: Reasons in a World of Causes*. Cambridge, MA: MIT Press.

Egan, A. 2007. "Some Counterexamples to Causal Decision Theory." *Philosophical Review* 116: 93–114.

Egré, P. and M. Cozic. 2010. "If-Clauses and Probability Operators." *Cahiers de Recherches, série Décision, Rationalité, Interaction*, Cahier DRI 2010–05. Paris: IHPST Éditions.

Elga, A. 2010. "Subjective Probabilities Should Be Sharp." *Philosophers' Imprint* 10, 5: 1–11. Ann Arbor, MI: University of Michigan. www.philosophersimprint. org/010005/

Feldman, F. 2000. "Basic Intrinsic Value." *Philosophical Studies* 99: 319–46.

Fine, K. 2012. "Counterfactuals without Possible Worlds." *Journal of Philosophy* 109: 221–46.

Freedman, D. 1997. "From Association to Causation via Regression." In V. McKim and S. Turner, eds., *Causality in Crisis?: Statistical Methods and the Search for Causal Knowledge in the Social Sciences*, pp. 113–61. Notre Dame, IN: University of Notre Dame Press.

Fuchs, A. 1985. "Rationality and Future Desires." *Australasian Journal of Philosophy* 63: 479–84.

Gibbard, A. 1986. "Interpersonal Comparisons: Preference, Good, and the Intrinsic Reward of a Life." In J. Elster and A. Hylland, eds., *Foundations of Social Choice Theory*, pp. 165–94. Cambridge: Cambridge University Press.

[1972] 1990. *Utilitarianism and Coordination*. New York: Garland Publishing. Publication of a Ph.D. dissertation.

2008. "Causal Influence and Controlled Experiment." Fifth Annual Formal Epistemology Workshop. http://fitelson.org/few/few_08/gibbard.pdf

Gibbard, A. and W. Harper. [1978] 1981. "Counterfactuals and Two Kinds of Expected Utility." In W. Harper, R. Stalnaker, and G. Pearce, eds., *Ifs: Conditional, Belief, Decision, Chance, and Time*, pp. 153–90. Dordrecht: Reidel.

Gigerenzer, G. 2000. *Adaptive Thinking: Rationality in the Real World*. New York: Oxford University Press.

Gilboa, I. 2009. *Theory of Decision under Uncertainty*. Cambridge: Cambridge University Press.

Gillies, A. 2007. "Counterfactual Scorekeeping." *Linguistics and Philosophy* 30: 329–60.

Gollier, C. 2001. *The Economics of Risk and Time*. Cambridge, MA: MIT Press.

Gorman, W. M. 1995. *Separability and Aggregation. Collected Works of W. M. Gorman*, Vol. 1. C. Blackorby and A. F. Shorrocks, eds. Oxford: Clarendon Press.

Hall, N. 2004. "Two Concepts of Causation." In J. Collins, E. J. Hall, and L. A. Paul, eds., *Causation and Counterfactuals*, pp. 225–76. Cambridge, MA: MIT Press.

Halpern, J. and J. Pearl. 2005. "Causes and Explanations: A Structural-Model Approach – Part 1, Causes." *British Journal for Philosophy of Science* 56: 843–87.

Hammond, P. 1988. "Consequentialist Foundations for Expected Utility." *Theory and Decision* 25: 25–78.

Hansson, S. O. 2001. *The Structure of Value and Norms*. Cambridge: Cambridge University Press.

Hanushek, E. and J. Jackson. 1977. *Statistical Methods for Social Scientists*. New York: Academic Press.

Harman, G. 1967. "Toward a Theory of Intrinsic Value." *Journal of Philosophy* 64: 792–804.

Hausman, D. 1998. *Causal Asymmetries*. Cambridge: Cambridge University Press.
   2012. *Preference, Value, Choice, and Welfare*. Cambridge: Cambridge University Press.

Hedden, B. 2012. "Options and the Subjective Ought." *Philosophical Studies* 158: 343–60.

Hitchcock, C. 2001. "The Intransitivity of Causation Revealed in Equations and Graphs." *Journal of Philosophy* 98: 273–99.

Hitchcock, C. and J. Kobe. 2009. "Cause and Norm." *Journal of Philosophy* 106: 587–612.

Horty, J. 2001. *Agency and Deontic Logic*. New York: Oxford.

Huang, J. and J. Bargh. 2014, "The Selfish Goal: Autonomously Operating Motivational Structures as the Proximate Cause of Human Judgment and Behavior." *Behavioral and Brain Sciences* 37: 121–35.

Hurka, T. 1998. "Two Kinds of Organic Unity." *Journal of Ethics* 2: 299–320.

Jeffrey, R. [1965] 1990. *The Logic of Decision*. Second edition, paperback. Chicago: University of Chicago Press.

Joyce, J. 1999. *The Foundations of Causal Decision Theory*. Cambridge: Cambridge University Press.
   2003. "Review of Paul Weirich's *Decision Space: Multidimensional Utility Analysis*." *Ethics* 113: 914–19.
   2010. "Causal Reasoning and Backtracking." *Philosophical Studies* 147: 139–54.

Kahneman, D. and A. Tversky. 1979. "Prospect Theory." *Econometrica* 47: 263–91.

Kahneman, D., E. Diener, and N. Schwarz, eds. 2003. *Well-Being: Foundations of Hedonic Psychology*. New York: Russell Sage Foundation.

Keeney, R. and H. Raiffa. [1976] 1993. *Decisions with Multiple Objectives: Preferences and Value Tradeoffs*. Second edition. Cambridge: Cambridge University Press.

Korcz, K. 2002. "The Epistemic Basing Relation." In E. Zalta, ed., *The Stanford Encyclopedia of Philosophy*. http://plato.stanford.edu/entries/basing-epistemic/

Korsgaard, C. 1983. "Two Distinctions in Goodness." *Philosophical Review* 92: 169–95.

Krantz, D., R. Luce, P. Suppes, and A. Tversky. 1971. *Foundations of Measurement*, Vol. 1. New York: Academic Press.

Kratzer, A. 2012. *Modals and Conditionals*. New York: Oxford University Press.

Kutach, D. 2012. *Causation and Its Basis in Fundamental Physics*. Oxford: Oxford University Press.

Kyburg, H. 1974. *The Logical Foundations of Statistical Inference*. Dordrecht: Reidel.

Leitgeb, H. 2012. "A Probabilistic Semantics for Counterfactuals." *Review of Symbolic Logic*, 5: 26–121.

Lemos, N. 1994. *Intrinsic Value: Concept and Warrant*. Cambridge: Cambridge University Press.

Lewis, D. 1973. *Counterfactuals*. Oxford: Blackwell.

1976. "Probabilities of Conditionals and Conditional Probabilities." *Philosophical Review* 85: 297–315.

[1973] 1986a. "Causation." In D. Lewis, *Philosophical Papers*, Vol. II, pp. 159–213. New York: Oxford University Press.

1986b. *On the Plurality of Worlds*. Malden, MA: Blackwell.

Loewenstein, G. and D. Schkade. 1999. "Wouldn't It Be Nice? Predicting Future Feelings." In D. Kahneman, E. Diener, and N. Schwarz, eds., *Well-Being: The Foundations of Hedonic Psychology*, pp. 85–105. New York: Russell Sage Foundation.

Malpezzi, S. 2003. "Hedonic Pricing Models: A Selective and Applied Review." In T. O'Sullivan and K. Gibb, eds., *Housing Economics and Public Policy*, pp. 67–89. Malden, MA: Blackwell Science.

Maslen, C. 2004. "Causes, Contrasts, and the Nontransitivity of Causation." In J. Collins, E. J. Hall, and L. A. Paul, eds., *Causation and Counterfactuals*, pp. 341–58. Cambridge, MA: MIT Press.

McCain, K. 2012. "The Interventionist Account of Causation and the Basing Relation." *Philosophical Studies* 159: 357–82. DOI:10.1007/s11098-011-9712-7.

McClennen, E. 1990. *Rationality and Dynamic Choice*. Cambridge: Cambridge University Press.

Meacham, C. and J. Weisberg. 2011. "Representation Theorems and the Foundations of Decision Theory." *Australasian Journal of Philosophy* 89: 641–63.

Mele, A. 2003. *Motivation and Agency*. New York: Oxford University Press.

Moore, G. E. [1903] 1993. *Principia Ethica*. Revised edition. Cambridge: Cambridge University Press.

Nelson, P. 1999. "Multiattribute Utility Models." In P. Earl and S. Kemp, eds., *The Elgar Companion to Consumer Research and Economic Psychology*, pp. 392–400. Cheltenham, UK: Edward Elgar.

Papineau, D. 1992. "Can We Reduce Causal Direction to Probabilities?" *PSA: Proceedings of the Biennial Meeting of the Philosophy of Science Association. Vol. 2: Symposia and Invited Papers*, pp. 238–52.

Parfit, D. 1984. *Reasons and Persons*. Oxford: Oxford University Press.

1986. "Overpopulation and the Quality of Life." In P. Singer, ed., *Applied Ethics*, pp. 145–64. Oxford: Oxford University Press.

Pearl, J. 2000. *Causality: Models, Reasoning, and Inference*. Cambridge: Cambridge University Press. Second edition, 2009.

Peterson, M. 2009. *An Introduction to Decision Theory*. Cambridge: Cambridge University Press.

Pettit, P. 2002. *Rules, Reasons, and Norms*. Oxford: Oxford University Press.

Pust, J. 2012. "Intuition." In E. Zalta, ed., *Stanford Encyclopedia of Philosophy*. http://plato.stanford.edu/archives/win2012/entries/intuition/

Quinn, W. 1974. "Theories of Intrinsic Value." *American Philosophical Quarterly* 11: 123–32.

Rabinowicz, W. and T. Rønnow-Rasmussen. 2000. "A Distinction in Value: Intrinsic and for Its Own Sake." *Proceedings of the Aristotelian Society* 100: 33–51.

Raibley, J. 2012. "Welfare over Time and the Case for Holism." *Philosophical Papers* 41: 239–65.

Ramsey, F. 1931. "Truth and Probability." In R. Braithwaite, ed., *The Foundations of Mathematics*, pp. 156–98. New York: Harcourt.

Robinson, M. and G. Clore. 2002. "Belief and Feeling: Evidence for an Accessibility Model of Emotional Self-Report." *Psychological Bulletin* 128: 934–60.

Rosen, S. 1974. "Hedonic Prices and Implicit Markets: Product Differentiation in Pure Competition." *Journal of Political Economy* 82: 34–55.

Rothschild, D. 2013. "Do Indicative Conditionals Express Propositions?" *Noûs* 47: 49–68.

Samuelson, L. 1997. *Evolutionary Games and Equilibrium Selection*. Cambridge, MA: MIT Press.

Sartorio, C. 2013. "Making a Difference in a Deterministic World." *Philosophical Review* 122: 189–214.

Savage, L. [1954] 1972. *The Foundations of Statistics*. Second edition. New York: Dover.

Schervish, M., T. Seidenfeld, and J. Kadane. 1990. "State-Dependent Utilities." *Journal of the American Statistical Association* 85, 411: 840–47.

Schroeder, T. 2004. *Three Faces of Desire*. New York: Oxford University Press.

Seidenfeld, T. 1988. "Decision Theory without 'Independence' or without 'Ordering': What Is the Difference?" *Economics and Philosophy* 4: 267–90.

Shafir, E., I. Simonson, and A. Tversky. 1993. "Reason-Based Choice." *Cognition* 49: 11–36.

Shanahan, M. 2009. "The Frame Problem." In E. Zalta (ed.), *Stanford Encyclopedia of Philosophy*. http://plato.stanford.edu/archives/win2009/entries/frame-problem/

Sloman, S. 2005. *Causal Models: How People Think about the World and Its Alternatives*. New York: Oxford University Press.

Sobel, J. H. 1994. *Taking Chances: Essays on Rational Choice*. Cambridge: Cambridge University Press.

Sober, E. 1988. "The Principle of the Common Cause." In J. Fetzer, ed., *Probability and Causality: Essays in Honour of Wesley Salmon*, pp. 211–29. Dordrecht: Reidel.

Sober, E. and D. Wilson. 1998. *Unto Others: The Evolution and Psychology of Unselfish Behavior*. Cambridge, MA: Harvard University Press.

Spirtes, P., C. Glymour, and R. Scheines. 2000. *Causation, Prediction, and Search*. Cambridge, MA: MIT Press.

Spohn, W. 2001. "Bayesian Nets Are All There Is to Causal Dependence." In M. Galavotti, P. Suppes, and D. Costantini, eds., *Stochastic Causality*, pp. 157–72. Stanford: CSLI Publications.

Stalnaker, R. [1968] 1981. "A Theory of Conditionals." In W. Harper, R. Stalnaker, and G. Pearce, eds., *Ifs*, pp. 41–55. Dordrecht: Reidel.

Taylor, R. 1962. "Fatalism." *Philosophical Review* 71: 56–66.

Varian, H. 1984. *Microeconomic Analysis*. Second edition. New York: Norton.

Velleman, J. D. 2000. "Well-Being and Time." In J. D. Velleman, ed., *Possibility of Practical Reason*, pp. 56–84. Oxford: Oxford University Press.

Weirich, P. 1980. "Conditional Utility and Its Place in Decision Theory." *Journal of Philosophy* 77: 702–15.

1981. "A Bias of Rationality." *Australasian Journal of Philosophy* 59: 31–37.

1986. "Expected Utility and Risk." *British Journal for the Philosophy of Science* 37: 419–42.

2001. *Decision Space: Multidimensional Utility Analysis*. Cambridge: Cambridge University Press.

2004. *Realistic Decision Theory: Rules for Nonideal Agents in Nonideal Circumstances*. New York: Oxford University Press.

2009. "Intrinsic Utility." In L. Johansson, J. Österberg, and R. Sliwinski, eds., *Logic, Ethics, and All That Jazz: Essays in Honour of Jordan Howard Sobel*, pp. 373–86. Uppsala, Sweden: Department of Philosophy, Uppsala University.

2010a. *Collective Rationality: Equilibrium in Cooperative Games*. New York: Oxford University Press.

2010b. "Utility and Framing." In P. Weirich, ed., *Realistic Standards for Decisions*, a special issue of *Synthese*, 176: 83–103. DOI:10.1007/s11229-009-9485-0.

2012. "Causal Decision Theory." In E. Zalta, ed., *Stanford Encyclopedia of Philosophy*. http://plato.stanford.edu/entries/decision-causal/

Westerståhl, D. and P. Pagin. 2011. "Compositionality." In C. Maienborn, K. von Heusinger, and P. Portner, eds., *Semantics: An International Handbook of Natural Language Meaning*, Vol. 1, pp. 96–123. Berlin: De Gruyter Mouton.

Woodward, J. 1997. "Causal Models, Probabilities, and Invariance." In V. McKim and S. Turner, eds., *Causality in Crisis? Statistical Methods and the Search for Causal Knowledge in the Social Sciences*, pp. 265–315. Notre Dame, IN: University of Notre Dame Press.

2003. *Making Things Happen: A Theory of Causal Explanation*. New York: Oxford University Press.

Zhang, J. 2012. "A Lewisian Logic of Causal Counterfactuals." *Mind and Machines* 23: 77–93.

2013. "A Comparison of Three Occam's Razors for Markovian Causal Models." *British Journal for the Philosophy of Science* 64: 423–48.

Zimmerman, M. 2001. *The Nature of Intrinsic Value*. Lanham, MD: Rowman and Littlefield.

Index

Printed in the United States
By Bookmasters

Printed in the United States
By Bookmasters